Problems of mixed mode crack propagation

ENGINEERING APPLICATION OF FRACTURE MECHANICS

Editor-in-Chief: George C. Sih

Also in this series:

G.C. Sih and L. Faria (eds.)
Fracture mechanics methodology
1984. ISBN 90 247 2941 6

Problems of mixed mode crack propagation

Emmanuel E. Gdoutos

School of Engineering,
Democritus University of Thrace,
Xanthi, Greece

1984 **MARTINUS NIJHOFF PUBLISHERS**
a member of the KLUWER ACADEMIC PUBLISHERS GROUP
THE HAGUE / BOSTON / LANCASTER

Distributors

for the United States and Canada: Kluwer Boston, Inc., 190 Old Derby Street, Hingham, MA 02043, USA
for all other countries: Kluwer Academic Publishers Group, Distribution Center, P.O.Box 322, 3300 AH Dordrecht, The Netherlands

Library of Congress Cataloging in Publication Data

Gdoutos, E. E., 1948-
Problems of mixed mode crack propagation.

(Engineering application of fracture mechanics ; v. 2)
Includes bibliographical references and indexes.
1. Fracture mechanics. I. Sih, G. C. (George C.)
II. Title. III. Series.
TA409.G383 1984 620.1'126 83-23626
ISBN-13: 978-94-009-6191-3

ISBN-13: 978-94-009-6191-3 e-ISBN-13: 978-94-009-6189-0
DOI: 10.1007/978-94-009-6189-0

Contents

Series on engineering application of fracture mechanics

Fracture mechanics technology has received considerable attention in recent years and has advanced to the stage where it can be employed in engineering design to prevent against the brittle fracture of high-strength materials and highly constrained structures. While research continued in an attempt to extend the basic concept to the lower strength and higher toughness materials, the technology advanced rapidly to establish material specifications, design rules, quality control and inspection standards, code requirements, and regulations for safe operation. Among these are the fracture toughness testing procedures of the American Society of Testing Materials (ASTM), the American Society of Mechanical Engineers (ASME) Boiler and Pressure Vessel Codes for the design of nuclear reactor components, etc. Step-by-step fracture detection and prevention procedures are also being developed by the industry, government and university to guide and regulate the design of engineering products. This involves the interaction of individuals from the different sectors of the society that often presents a problem in communication. The transfer of new research findings to the users is now becoming a slow, tedious and costly process.

One of the practical objectives of this series on *Engineering Application of Fracture Mechanics* is to provide a vehicle for presenting the experience of real situations by those who have been involved in applying the basic knowledge of fracture mechanics in practice. It is time that the subject should be presented in a systematic way to the practicing engineers as well as to the students in universities at least to all those who are likely to bear a responsibility for safe and economic design. Even though the current theory of linear elastic fracture mechanics (LEFM) is limited to brittle fracture behavior, it has already provided a remarkable improvement over the conventional methods not accounting for initial defects that are inevitably present in all materials and structures. The potential of the fracture mechanics technology, however, has not been fully recognized. There remains much to be done in constructing a quantitative theory of material damage that can reliably translate small specimen data to the design of large size structural components. The work of the physical metallurgists and the fracture mechanicians should also be brought together by reconciling the details of the material microstructure with the assumed continua of the computational methods. It is with the aim of developing a wider appreciation of the fracture mechanics technology applied to the design of engineering structures such as aircrafts, ships, bridges, pavements, pressure vessels, off-shore structures, pipelines, etc. that this series is being developed.

Undoubtedly, the successful application of any technology must rely on the soundness of the underlying basic concepts and mathematical models and how they reconcile with each other. This goal has been accomplished to a large extent by the book series on *Mechanics of Fracture* started in 1972. The seven published volumes offer a wealth of information on the effects of defects or cracks in cylindrical bars, thin and thick plates, shells, composites and solids in three dimensions. Both static and dynamic loads are considered. Each volume contains an introductory chapter that illustrates how the strain energy criterion can be used to analyze the combined influence of defect size, component geometry and size, loading, material properties, etc. The criterion is particularly effective for treating mixed mode fracture where the crack propagates in a non-self similar fashion. One of the major difficulties that continuously perplex the practitioners in fracture mechanics is the selection of an appropriate fracture criterion without which no reliable prediction of failure could be made. This requires much discernment, judgement and experience. General conclusion based on the agreement of theory and experiment for a limited number of physical phenomena should be avoided.

Looking into the future the rapid advancement of modern technology will require more sophisticated concepts in design. The micro-chips used widely in electronics and advanced composites developed for aerospace applications are just some of the more well-known examples. The more efficient use of materials in previously unexperienced environments is no doubt needed. Fracture mechanics should be extended beyond the range of LEFM. To be better understood is the entire process of material damage that includes crack initiation, slow growth and eventual termination by fast crack propagation. Material behavior characterized from the uniaxial tensile tests must be related to more complicated stress states. These difficulties could be overcome by unifying metallurgical and fracture mechanics studies, particularly in assessing the results with consistency.

This series is therefore offered to emphasize the applications of fracture mechanics technology that could be employed to assure the safe behavior of engineering products and structures. Unexpected failures may or may not be critical in themselves but they can often be annoying, time-wasting and discrediting of the technical community.

Bethlehem, Pennsylvania G.C. Sih
1984 Editor-in-Chief

Editor-in-Chief's preface

This is not just another book on fracture mechanics. In recent years, there have been many books published on this subject in an attempt to assess the state of the art and its applications. The majority of the work dealt with energy release rate or critical stress intensity factor and is applicable only to fracture toughness testing. The main reason for this restriction is that the energy release concept cannot easily be extended to mixed mode fracture that occurs in practice as the rule rather than the exception. Cracks will normally curve or turn because the direction of loading can change as a function of time. Their directions of growth cannot be assumed as an a priori and must be determined from a pre-assumed criterion. Analysts are still perplexed with selecting an appropriate fracture criterion because it requires much discernment and judgement. Criteria which often appeared valid for idealized situations are quickly discredited when encountering more complex physical phenomena. Moreover, the claim of generality cannot be justified on the basis of agreement between theory and experiment for a few simple examples.

From a practical viewpoint the need for a fracture criterion that can treat real situations is apparent. The strain energy density criterion has been applied successfully, not only for complex loading and geometric conditions encountered in service, but also for materials in the elastic, elastic–plastic and fully plastic range. The generality of this criterion has been demonstrated in many applications including ductile and fatigue crack growth. It is time that a systematic treatment of the criterion be presented to illustrate the versatility of the strain energy density concept. The *allowable load* and *component size* for a variety of complex crack configurations and loadings are obtained. The book can be used in undergraduate and postgraduate courses, by all engineers in the making and by all those who are interested in developing design methodologies to include the influence of initial defects or cracks. It is from detailed consideration of many of the example problems provided in the book that proper design codes and methods will evolve and the further improvement of fracture mechanics technology will develop.

The book offers a degree of consistency for predicting the onset of rapid fracture for many crack systems never available before. Readers are therefore encouraged to employ the results to good practical use in designing structural components relative to the tolerable defect or crack sizes.

Bethlehem, Pennsylvania
1984

G.C. Sih
Editor-in-Chief

Author's preface

Fracture mechanics has received much attention in the last two decades for determining the useful life of structural members. This discipline is based on the principle that all materials contain initial defects in the form of cracks, voids or inclusions which can affect the load-carrying capacity of engineering structures. The quantitative assessment of these defects requires a complete knowledge of the local stress and strain fields coupled with a suitable fracture criterion. Although considerable progress has been made in evaluating the stresses and strains in solids weakened by cracks for a variety of geometries and loading conditions, the opinions on selecting the appropriate failure criteria remain diversified. This is mainly because too much emphasis has been placed merely on the fitting of experimental data with analytical prediction from one special situation to another. A unified approach to couple stress and failure analysis is lacking.

The usefulness of a failure criterion should be tested by its consistency and ability to explain the physical phenomena in general. Many of the more conventional criteria, such as maximum normal stress, critical stress intensity factor, or energy release rate, etc., have had some limited success in analyzing the simpler crack problems, but are found to be inadequate and often invalid for the more complex situations. Their limitations lie not only in the elasticity and linearity of material or structure behavior but also in dimensionability and symmetry between load and crack orientation. Unverified assumptions that were thought to yield conservative results are frequently invalidated when applied to situations in practice.

More specifically, the shortcomings of the above mentioned fracture criteria are their restrictions to symmetry between the applied load and crack plane and self-similar crack growth. In other words, the direction and shape of crack growth must be known as an a priori. Such an idealization is seldom encountered in service and must be regarded as the exception rather than the rule. In addition to the treatment of curved crack path, the resistance of the material to fracture should also be included in the theory without additional assumptions.* This property is manifested through the fracture toughness parameter which is intended to be characteristic of the material regardless of whether the material deforms, elastically and/or plastically. The ability of a criterion to treat mixed mode fracture cannot be overemphasized. Mixed mode fracture can occur in the plane of the crack specimen when load and crack are not symmetrically aligned or in the thickness direction when ductile fracture modes are present such as cup-and-cone failure or the development of shear lips near the specimen surfaces.

*Although the maximum circumferential stress criterion has been frequently used to determine the direction of crack initiation, the critical stress quantity is not representative of the fracture toughness of the material.

In the early 1970's, Sih proposed a fracture criterion that fundamentally represents a departure from the conventionally accepted criteria mentioned previously. This new criterion does not rely on the existence of an initial defect such as a crack in the solid. It can account for failure of the material by fracture and/or yielding. This permits a consistent treatment of the entire failure process starting from the early stage of fracture initiation to the final separation of the material. One of the basic ideas is that failure initiation is assumed to occur when material elements break. Since the exact location cannot be determined by experiments, it necessitates the introduction of a small length parameter to define a finite region. This parameter is often referred to as the radius of the core region within which the continuum mechanics analysis is no longer valid.

Many experiments have been carried out to establish the validity of this region. Hence, the elements under consideration are assumed to lie exterior to the core region and their angular positions at fracture initiation are determined by the relative minima of the strain energy density function. At the site of the fracture, the dilatational component of the strain energy density function is dominant in comparison with the distortional component as the former can be associated with volume change leading to the creation of new free surfaces and the latter with yielding. The proportion of these components is weighed automatically by taking the minimum values of the strain energy density function with respect to the reference coordinate. The direction of maximum yielding corresponds to the relative maximum of the strain energy density. The critical value of the strain energy density to initiate yielding and fracture are obviously different. It is well known in the case of a macrocrack stretched symmetrically, yielding occurs to the side of the crack while fracture takes place straight ahead. Yielding and fracture occur at *different locations*. Hence, there should be no confusion with reference to the physical meaning of the relative maximum and minimum values of the strain energy density function.

The *strain energy density criterion* will be applied throughout this book for determining the allowable load corresponding to crack initiation for a variety of engineering problems of practical interest. The combined effects of crack geometry and loading can be expressed in terms of the critical strain energy density factor for a given material. This critical parameter represents the fracture toughness of the material and has been determined experimentally for many structural engineering materials. The problem statement and results are given in nine chapters.

The first chapter provides an introduction of the strain energy density theory applied to linear elastic materials. Addressed in the second chapter are crack problems in two-dimensions. The strain energy density expressions are established in terms of the stress intensity factors that describe the intensity of the crack tip stress field. Treated are the problems of a single crack inclined to a uniform applied stress and of an infinite plate loaded by concentrated forces and/or moments. The third chapter considers the interaction of different types of branched cracks under simple loading conditions. Both symmetrically and asymmetrically branched cracks are considered. The angle crack and crack branches from an elliptical cavity are also investigated. The fourth chapter is devoted to plates with interacting cracks of various geometrical configurations. Some of the configurations considered are two equal skew parallel cracks, two equal symmetrically or arbitrarily oriented cracks, etc. Arc-shaped cracks are investigated in the fifth chapter. The single crack and an array of star-shaped circular cracks are considered. The sixth chapter considers the cases where cracks are emanated from holes or inclusions. Included are perforated plates with one, two or an array of equal cracks. The problem of failure initiating from singular points of rigid inclusions is solved. Results are also given for the hypocycloidal inclusion and a rigid line inclusion. This leads to the topic of composite materials in chapter seven. The case of a bimaterial plate with an arbitrarily oriented crack to the interface, the three-layered composite with cracks, the bimaterial plate with a circular crack, and bimaterial plate with a crack along the interface and the interaction between a crack and a circular inclusion are among the composite systems analyzed.

While the first seven chapters deal with two-dimensional crack problems of various crack configurations and loading conditions, the remaining two chapters are devoted to three-dimensional crack problems. Chapter eight considers the cases of a bent plate and shell with cracks. The bent plate contains a through crack loaded by a bending moment in an arbitrary direction. The cracked shell represents a pressurized long cylindrical vessel with two semispherical heads. Finally, chapter nine presents the problem of a flat elliptical crack embedded in a solid loaded by a uniform inclined stress.

It is not the intent of this book to overwhelm the readers with the vast number of example problems, but rather to demonstrate the usefulness and versatility of the strain energy density criterion. Many of the situations considered cannot be analyzed by the conventional criteria. Although most of the crack problems treated are in two-dimensions, the criterion applies equally well to three dimensions and other more complicated loading conditions.

The manuscript was completed while the author was on sabbatical leave from the Democritus University of Thrace of Greece to the Institute of Fracture and Solid Mechanics at Lehigh University. The author wishes to express his gratitude to Professor G.C. Sih for his very stimulating discussions, his comments and suggestions during the writing of the book. Thanks are also due to the Institute Secretary, Mrs. Constance Weaver, for typing the manuscript and other assistance she provided during my stay at Lehigh.

Xanthi, Greece, 1984 E.E. Gdoutos

1

The strain energy density criterion

1.1. Introductory remarks

It is a common experience that breakage of a material under load can be preceded either by large or small deformation depending on the temperature, rate of loading and other variables present. The former occurs over a relatively long time period and may be associated with yielding or plastic flow. The criterion of Tresca or Mises may be used to describe this process. The latter is concerned with breaking suddenly and is accompanied by little deformation. This process is usually referred to as brittle fracture.

It was demonstrated that the strength of a material measured from a laboratory specimen is many times lower than that predicted from theoretical calculations depending on the state of affairs internally. Experimental studies reveal that all materials contain mechanical defects in the form of cracks, voids or inclusions which are not always accounted for in the analysis. Near the defects, high stresses prevail that are often responsible for lowering the strength of the material. Study on the load-carrying capacity of structures containing initial flaws gave impetus to the development of fracture mechanics. One of the objectives of this discipline is to determine the critical load of a structure by accounting for the size and location of initial defects. For those materials that do not deform appreciably prior to fracture, the theory of linear elasticity can be used to describe the state of stress and deformation in the presence of defects or cracks. This assumption forms the basis of the linear elastic fracture mechanics theory that plays an important role in the prediction of catastrophic failure. It is therefore desirable to find a parameter which can account for the resistance of a material against the initiation of crack extension so that the sensitivity of materials to fracture can be better understood. In addition, it is also important to have the ability to predict crack trajectory or fracture path before and after material separation. Such a knowledge is valuable in engineering design.

The most widely used fracture criterion for predicting the brittle fracture of solids is that of Griffith [1, 2]. Based on the concept of the energy release, this criterion states that a crack will begin to propagate when the elastic energy released by a small increment of growth is greater than the energy required to create the fractured surfaces. For a crack of length $2a$ in a plate subjected to a uniform stress σ applied normal to the crack, it is found that the critical stress σ_{cr} corresponding to the onset of crack extension is

$$\sigma_{cr}a^{1/2} = \left[\frac{2E\gamma}{\pi(1-\nu^2)}\right]^{1/2} = \text{material constant} \tag{1.1}$$

where the plane strain condition is assumed. In equation (1.1), E is the elastic modulus,

ν the Poisson's ratio and γ the specific surface energy. Since the quantity in the right hand side of equation (1.1) depends only on material constants, the factor $\sigma_{cr}a^{1/2}$ may be regarded as an intrinsic material parameter. This was verified experimentally by Griffith in tests on glass specimens containing cracks for a wide range of different lengths.

The Griffith theory of fracture, however, has two main limitations. The first is associated with the implicit assumption that the crack extends continuously into the material. This idealized assumption is contrary to experimental observation. Crack extension has been known as a discontinuous process. Crack grows in increments according to load increments that are generally nonuniform. The second and more serious limitation is that Griffith assumes the direction of crack growth to be known as an a priori. Hence, equation (1.1) applies only when the load is applied normal to the crack such that the crack propagates in a self-similar manner. However, in practice nonalignment of loads with respect to crack orientation is a common practice. In such cases, the crack-tip stress field is no longer governed by the opening-mode stress intensity factor k_1 but involves the combination of three stress intensity factors k_1, k_2, and k_3. Moreover, the direction of crack initiation is not known as an a priori and depends on some combination of k_1, k_2, and k_3 satisfying certain failure criterion.

Prediction on crack instability due to mixed mode loading conditions has been made with two basic viewpoints. The Griffith energy release rate approach has been extended to the analysis of a small segment of crack extension tilted at an angle with respect to the original crack plane. Besides the mathematical complexities involved in the analysis of the branched crack problem, the uniqueness of the energy release rate quantity has been questioned [3–14] when the tilted portion of the crack tends to zero in the limit. This involves the overlapping of two singular points with different orders of singularity whose limit may or may not be unique. No proof exists. The second viewpoint considers the stress state corresponding to the initial crack configuration for predicting crack extension. It has been shown experimentally [15] that material elements at a finite distance ahead of the crack fail first before the onset of crack extension. This led to the search of a parameter associated with the failure of a material element ahead of the crack. The circumferential stress σ_θ was first employed [16]. This criterion assumes that the onset of crack extension corresponds to σ_θ reaching some maximum value, $(\sigma_\theta)_{max}$, at a finite distance from the crack tip [17, 18]. The criterion predicted angles of crack initiation with reasonable accuracy. The results were confirmed by experiments on plexiglass plates with inclined cracks subjected to remote tension. Although the maximum stress criterion can predict the direction of crack initiation for some mixed mode loading conditions, it is inadequate for a number of reasons. First, the location of failure may not always be governed by only one of the six independent stress components. The combination may play a role. Next, a stress quantity cannot be used to describe the fracture resistance of a material. An even more serious objection is the contradiction that occurs for the case of a moving crack where the normal stress parallel to the crack is greater than the stress perpendicular to the crack. This example shows that the maximum stress criterion is in direct disagreement with experimental observation.

The need of a criterion that can predict failure in a simple and unified manner has led to the development of the strain energy density theory [19–22]. A most important feature of a useful criterion is that it must be free from physical contractions and sufficiently general for treating engineering problems encountered in practice. The strain energy density criterion focuses attention on an element of the material ahead of the crack. Failure is then assumed to be governed by the stationary values of the strain energy density function. The basic principles of this criterion will be discussed in the sections to follow.

1.2. The strain energy density concept

Generally speaking, description of failure of a material element by yielding or fracture requires the consideration of six stress and/or strain components. Two separate theories have been developed for the case of yielding. They are the total energy or Beltrami–Haigh theory and the distortional energy or Huber–Von Mises–Hencky theory. According to these theories, failure in a material by yielding occurs when the total or the distortional strain energy per unit volume absorbed by the material equals to the energy per unit volume stored in the material loaded in uniaxial tension at yield. This quantity corresponds to the limiting strain energy and is regarded as a material constant. Extensive experimental evidence is available on the use of the strain energy quantity to describe failure by yielding [23, 24].

A more generalized application of the local strain energy density in a material element at a possible site of failure has been proposed by Sih [19–22] as the dominant quantity which governs brittle fracture and/or yielding. The site of failure can be a crack tip, a notch front, a re-entrant point or a point of highly stressed material in an unflawed structure. This theory is developed on the basis of a strain energy density factor S for a material element at a finite distance r_0 from the point of failure initiation (Figure 1.1). Note that S is defined by

$$S = r_0 \frac{\mathrm{d}W}{\mathrm{d}V} \tag{1.2}$$

where $\mathrm{d}W/\mathrm{d}V$ is the strain energy density function per unit volume.

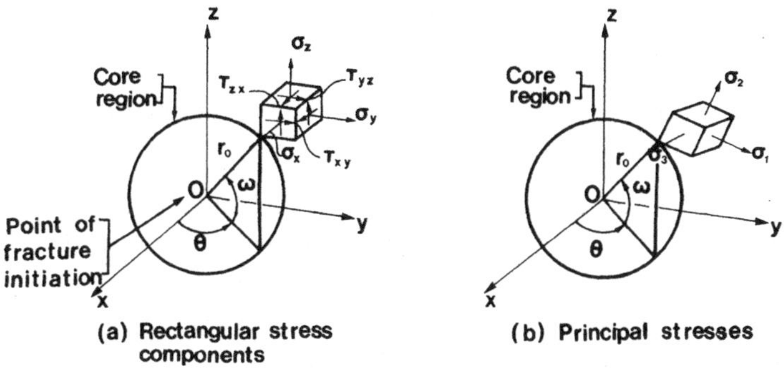

Figure 1.1. Spherical core region surrounding point 0 of failure initiation and a material element outside the core region. Rectangular stress components (a) and principal stresses (b).

When inelastic deformation is negligibly small everywhere in a solid, linear elastic fracture mechanics can be used to describe the state of stress and strain. Thus, the quantity $\mathrm{d}W/\mathrm{d}V$ can be written as

$$\frac{\mathrm{d}W}{\mathrm{d}V} = \frac{1}{2E}(\sigma_x^2 + \sigma_y^2 + \sigma_z^2) - \frac{\nu}{E}(\sigma_x\sigma_y + \sigma_y\sigma_z + \sigma_z\sigma_x) + \frac{1}{2\mu}(\tau_{xy}^2 + \tau_{yz}^2 + \tau_{zx}^2) \quad (1.3)$$

where σ_x, σ_y, σ_z, τ_{xy}, τ_{yz}, and τ_{zx} are the stress components, E the Young's modulus, ν the Poisson's ratio and μ the shear modulus of elasticity such that $E = 2\mu(1 + \nu)$.

For the plane problems of elasticity the quantity $\mathrm{d}W/\mathrm{d}V$ takes the form

$$\frac{\mathrm{d}W}{\mathrm{d}V} = \frac{1}{4\mu}\left[\frac{\kappa + 1}{4}(\sigma_x + \sigma_y)^2 - 2(\sigma_x\sigma_y - \tau_{xy}^2)\right] \quad (1.4)$$

where $\kappa = 3 - 4\nu$ for plane strain and $\kappa = (3 - \nu)/(1 + \nu)$ for generalized plane stress.

The strain energy per unit volume, $\mathrm{d}W/\mathrm{d}V$, can be further decomposed into two parts:

$$\frac{\mathrm{d}W}{\mathrm{d}V} = \left(\frac{\mathrm{d}W}{\mathrm{d}V}\right)_d + \left(\frac{\mathrm{d}W}{\mathrm{d}V}\right)_v \quad (1.5)$$

in which

$$\left(\frac{\mathrm{d}W}{\mathrm{d}V}\right)_d = \frac{1 + \nu}{6E}[(\sigma_x - \sigma_y)^2 + (\sigma_y - \sigma_z)^2 + (\sigma_z - \sigma_x)^2 + 6(\tau_{xy}^2 + \tau_{yz}^2 + \tau_{zx}^2)] \quad (1.6)$$

represents the distortional strain energy per unit volume corresponding to the deviatoric stress tensor that is associated with distortion of an element undergoing no volume change.

The quantity

$$\left(\frac{\mathrm{d}W}{\mathrm{d}V}\right)_v = \frac{1 - 2\nu}{6E}(\sigma_x + \sigma_y + \sigma_z)^2 \quad (1.7)$$

represents the part of the strain energy per unit volume associated with volume change and no shape change.

By means of the plane strain condition

$$\sigma_3 = \nu(\sigma_1 + \sigma_2) \quad (1.8)$$

where σ_1, σ_2, σ_3 are the principal stresses, the ratio $(\mathrm{d}W/\mathrm{d}V)_v/(\mathrm{d}W/\mathrm{d}V)_d$ obtained from equations (1.6) and (1.7) takes the form

$$\frac{(\mathrm{d}W/\mathrm{d}V)_v}{(\mathrm{d}W/\mathrm{d}V)_d} = \frac{(1 - 2\nu)(1 + \nu)[(\sigma_1/\sigma_2) + 1]^2}{[(\sigma_1/\sigma_2) - 1]^2 + [(1 - \nu) - \nu(\sigma_1/\sigma_2)]^2 + [(1 - \nu)(\sigma_1/\sigma_2) - \nu]^2} \quad (1.9)$$

Figure 1.2 gives the variations of the ratio $(\mathrm{d}W/\mathrm{d}V)_v/(\mathrm{d}W/\mathrm{d}V)_d$ with the ratio of the principal stresses σ_1/σ_2 for $\nu = 0$, 0.1, 0.2, 0.3, and 0.4. It is observed from this plot that the greatest volume change takes place for $\sigma_1 = \sigma_2$ corresponding to a two-dimensional hydrostatic stress state. For most metals with ν ranging from 0.2 to 0.3,

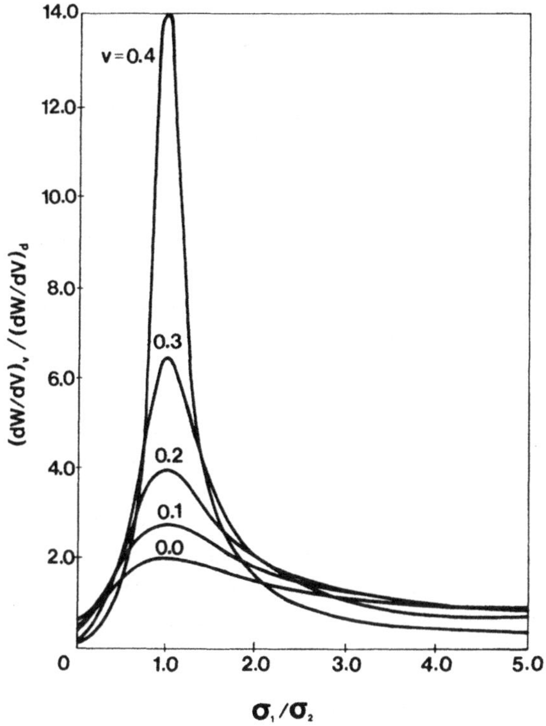

Figure 1.2. Variation of the ratio of the dilatational and distortional energy $(dW/dV)_v/(dW/dV)_d$ versus the ratio σ_1/σ_2 of the principal stresses for a state of plane strain. The Poisson's ratio ν takes the values 0, 0.1, 0.2, 0.3 and 0.4.

$(dW/dV)_v/(dW/dV)_d$ varies from 4 to 6.5. From the relative magnitudes of $(dW/dV)_v$ and $(dW/dV)_d$, it is seen that both quantities should be taken into account when considering the failure of material elements either by yielding* and/or fracture.

The introduction of a core region surrounding the failure site arises from the inability to describe in detail the state of stress and displacement in the immediate vicinity of the geometric discontinuity or free surface.** The radius of this core region r_0 when multiplied by dW/dV as shown in equation (1.2) gives the strain energy density factor S. Indeed, the analytical solution of the stress and displacement field based on theory of continuum mechanics must necessarily be defined at a finite distance away from the point where the influence of microstructure becomes dominant. Within the core region, the continuum mathematical model can no longer adequately describe the physical behavior of the material which is being highly strained and may even be inhomogeneous.

*The Huber–Von Mises–Hencky criterion of yielding neglects the portion of strain energy density due to volume change which cannot be justified in general.

**A limiting distance must be associated with all theories such that material damage is defined with reference to a given scale level, either microscopic and/or macroscopic.

The radius r_0 of the core region can be estimated approximately by assuming that the strain energy per unit volume in an element at distance r_0 from the point of fracture initiation takes its critical value determined from a simple tension test. An attempt for determining the size of this core region has been made by Theocaris [25, 26] who used the optical method of caustics. For most engineering metal alloys, r_0 was found to be of the order 10^{-3} to 10^{-2} in. Outside the core region, the stress field can adequately be described by the continuum mechanics solution. Hence, the boundary of the core region serves to separate the outside material assumed to behave elastically from the inside material whose mechanical properties are unknown. The size r_0 varies from material to material. Hence, it also reflects the material property. Sih and Kipp [18] have also made an estimate of r_0 on a theoretical basis from a study involving the radius of curvature of the crack tip. They analyzed the various crack trajectories emanating from notches of different sizes and noticed that at a finite distance from the notch tip the trajectories coincided with those for a sharp crack tip. Thus, the core region does not only depend on the intrinsic material properties but also on the local geometry of the mechanical imperfection. The estimates of Sih and Kipp for plexiglass agree remarkably well with those found by Theocaris [25, 26].

In accordance with what has been stated, the strain energy density factor S defined by equation (1.2) is to be calculated outside the core region. This approach is in agreement with the physical behavior of the material at failure as observed by experiments.

1.3. Basic hypotheses of the theory

To reiterate, the strain energy density theory proposed by Sih [19–22] can simultaneously account for fracture initiation and yielding for regions with different geometric configurations subjected to general loading conditions. The fundamental parameter used in the theory is the strain energy density factor S which is direction sensitive in that it predicts the direction of fracture initiation and/or propagation and yielding. The basic hypotheses of the theory apply to any type of body in three dimensions with or without inherent flaws and can be stated as follows:

Hypothesis (1): The direction of fracture initiation in a three-dimensional space is toward the point of minimum strain energy density factor, $S_{\min}$ as compared with other points on the same spherical surface surrounding the point.

Hypothesis (2): Fracture initiation is assumed to occur when $S_{\min}$ reaches the maximum critical value S_{cr}.

Hypothesis (3): The length r of the initial crack extension is assumed to be proportional to $S_{\min}$ such that $S_{\min}/r$ remains constant along the newly created fracture surface.

Consider a three-dimensional space in the rectangular Cartesian coordinate system x, y and z. A spherical core region of radius r_0 surrounds the point 0 that coincides with a geometric discontinuity such as a crack. Referring to Figure 1.1 the position of a material element is defined by the angles ω and θ. The third hypothesis arises from the fact that for three-dimensional crack problems r may vary from point to point on the crack border. The locations of fracture must be determined from a condition such as $S_{min}/r = (dW/dV)_{cr} =$ constant. For two-dimensional crack problems Hypotheses (1) and (2) are sufficient for determining the fracture path since the crack front is assumed to be straight in the thickness direction. It has been established experimentally that S_{cr} is a material constant [27, 28] and can account for the characterization of the fracture toughness of the material. Table 1.1 gives the values of S_{cr} together with the ultimate strength σ_u and the critical stress intensity factor K_c for some common metals and alloys.

Table 1.1. Fracture toughness values

Material	Ultimate strength σ_u (ksi)	Critical stress-intensity factor K_c (ksi$\sqrt{\text{in}}$)	Critical strain-energy-density factor S_{cr} (lb/in)†
A517F Steel (AM)	120	170	95.8
AISI 4130 Steel (AM)	170	100	33.2
AISI 4340 Steel (VAR)	300	40	5.3
AISI 4340 Steel (VAR)	280	40	5.3
AISI 4340 Steel (VAR)	260	45	6.7
AISI 4340 Steel (VAR)	240	60	11.9
AISI 4340 Steel (VAR)	220	75	18.7
300M Steel (VAR)	300	40	5.3
300M Steel (VAR)	280	40	5.3
300M Steel (VAR)	260	45	6.7
300M Steel (VAR)	240	60	11.9
300M Steel (VAR)	220	75	18.7
D6AC Steel (VAR)	240	40–90	5.3–26.9
H-11 Steel (VAR)	320	30	3.0
H-11 Steel (VAR)	300	40	5.3
H-11 Steel (VAR)	280	45	6.7
12Ni-5Cr-3Mo Steel (VAR)	190	220	160.5
18Ni (300) Maraging Steel (VAR)	290	50	8.3
18Ni (250) Maraging Steel (VAR)	260	85	24.0
18Ni (200) Maraging Steel (VAR)	210	120	47.8
18Ni (180) Maraging Steel (VAR)	195	160	84.9
9Ni-4Co-0.3C Steel (VAR)	260	60	11.9
Al 2014-T651	70	23	3.6
Al 2024-T851	65	23	3.6
Al 2219-T851	66	33	7.4
Al 2618-T651	64	32	7.0
Al 7001-T75	90	25	4.2
Al 7075-T651	83	26	4.6
Al 7079-T651	78	29	5.7
Al 7178-T651	83	24	3.9

†The calculation of S_c was based on $E = 30 \times 10^6$ psi, $\nu = 1/4$ for steel and $E = 10.6 \times 10^6$ psi. $\nu = 1/3$ for aluminum.

If the function $S(\theta, \omega)$ defined in equation (1.2) is differentiable in the vicinity of failure initiation, then Hypothesis (1) can be expressed mathematically by the relations

$$\frac{\partial S}{\partial \theta} = 0 \quad \text{and} \quad \frac{\partial S}{\partial \omega} = 0 \quad \text{at } (\theta_0, \omega_0) \tag{1.10}$$

$$\left(\frac{\partial^2 S}{\partial \theta \partial \omega}\right)^2 - \frac{\partial^2 S}{\partial \theta^2}\frac{\partial^2 S}{\partial \omega^2} < 0; \quad \frac{\partial^2 S}{\partial \theta^2} > 0 \text{ and } \frac{\partial^2 S}{\partial \omega^2} > 0 \quad \text{at } (\theta_0, \omega_0) \tag{1.11}$$

where (θ_0, ω_0) is the point along which failure is directed.

Relations (1.10) and (1.11) can be used for determining the point (θ_0, ω_0) at which the strain energy density factor S attains a relative minimum. When the function $S_{min} = S(\theta_0, \omega_0)$ reaches the value S_{cr} assumed to be a material constant, then fracture initiation will occur.

While the minimum value, S_{min}, of the strain energy density factor S is associated with brittle fracture, the maximum value S_{max} is related to yielding. The direction of S_{max} coincides with the location of maximum distortion, while S_{min} is connected with dilatation. Dilatation or volume change of material elements has been known to be related to the creation of free surface while distortion to yielding. This is illustrated in Figure 1.3 for the case of a crack under symmetric loading.

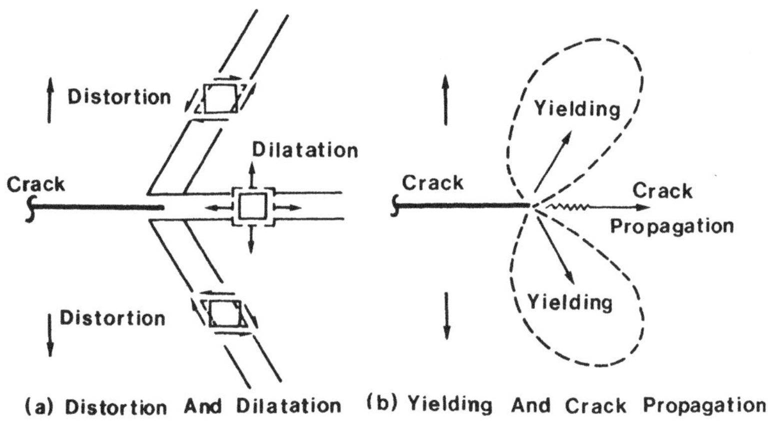

Figure 1.3. Dilatation and distortion of material elements ahead of a crack (a). Macrocrack propagation takes place in the direction of maximum dilatation and macroyielding in the direction of maximum distortion (b).

1.4. Fracture trajectories

Another salient feature of the strain energy density criterion is that it can be used not only for finding the critical load at failure but also for the direction of fracture initiation and the subsequent fracture path. For brittle fracture behavior, failure occurs suddenly and it does not seem unreasonable to assume that the fracture path can be determined from the conditions before fracture initiation takes place [29]. In such cases, the material does not have sufficient time to redistribute the stresses during the fracture process and the fracture path can be predetermined with high accuracy. A sound justification in favor of such behavior comes from the situation of brittle fracture of blunt notches or line cracks at low angle of loading where the loads required to initiate fracture are relatively high. In other words, the presence of a high density energy field near and far from the initial crack extension area are responsible for the rapid fracture when other physical resources exist to inhibit the fracture from propagating. The fracture trajectories and the corresponding critical loads can now be determined from the calculation of the fracture angles and loads for a wide range of radius vectors from the region of fracture initiation.

References

[1] Griffith, A.A., The phenomena of rupture and flow in solids, *Philosophical Transactions, Royal Society of London*, Series A221, pp. 163–198 (1921).

[2] Griffith, A.A., The theory of rupture, *Proceedings of First International Congress of Applied Mechanics*, Delft, pp. 55–93 (1924).

[3] Andersson, H., Stress-intensity factors at the tips of a star-shaped contour in an infinite tensile sheet, *Journal of the Mechanics and Physics of Solids*, 17, pp. 405–417 (1969).

[4] Andersson, H., Erratum on [3], *Journal of the Mechanics and Physics of Solids*, 18, p. 437 (1970).

[5] Palaniswamy, K. and Knauss, W.G., On the problem of crack extension in brittle solids under general loading, *Report SM 74-8, Graduate Aeronautical Lab.*, Caltech.

[6] Bilby, B.A. and Cardew, G.E., The crack with a kinked tip, *International Journal of Fracture*, 11, pp. 708–711 (1975).

[7] Khrapkov, A.A., The first basic problem for a notch at the apex of an infinite wedge, translated by I. Sneddon, *International Journal of Fracture Mechanics*, 7, pp. 373–382 (1971).

[8] Dudukalenko, V.V. and Romalis, N.B., Direction of crack growth under plane–stress–state conditions, *Izv. AN SSSR, Mekhanika Tverdogo Tela.*, 8, pp. 129–136 (1973).

[9] Hussain, M.A., Pu, S.L. and Underwood, G., Strain energy release rate for a crack under combined Mode I and Mode II, *Technical Report Benet Weapons Lab.*, August (1973).

[10] Theocaris, P.S., Assymetric branching of cracks, *Journal of Applied Mechanics*, 44, pp. 611–618 (1977).

[11] Nuismer, R.J., An energy release rate criterion for mixed mode fracture, *International Journal of Fracture*, 11, pp. 245–250 (1975).

[12] Chatterjee, S.N., The stress field in the neighborhood of a branched crack in an infinite elastic sheet, *International Journal of Solids and Structures*, 11, pp. 521–538 (1975).

[13] Vitek, V., Plane strain stress intensity factors for branched cracks, *International Journal of Fracture*, 13, pp. 481–501 (1977).

[14] Lo, K.K., Analysis of branched cracks, *Journal of Applied Mechanics*, 45, pp. 797–802 (1978).

[15] Sullivan, A.M., Effect of deformation upon the density of steel and copper, *M.S. Thesis*, University of Maryland (1955).

[16] Erdogan, F. and Sih, G.C., On the crack extension in plates under plane loading and transverse shear, *ASME Journal of Basic Engineering*, 85, pp. 519–527 (1963).

[17] Williams, J.G. and Ewing, P.D., Fracture under complex stress – the angled crack problem, *International Journal of Fracture Mechanics*, 8, pp. 441–446 (1972).

[18] Sih, G.C. and Kipp, M.E., Discussion on [17], *International Journal of Fracture*, 10, pp. 261–265 (1974).

[19] Sih, G.C., Some basic problems in fracture mechanics and new concepts, *Engineering Fracture Mechanics*, 5, pp. 365–377 (1973).

[20] Sih, G.C., Energy-density concept in fracture mechanics, *Engineering Fracture Mechanics*, 5, pp. 1037–1040 (1973).
[21] Sih, G.C., A special theory of crack propagation: methods of analysis and solutions of crack problems, *Mechanics of Fracture 1*, edited by G.C. Sih, Noordhoff International Publishing, Leyden, pp. XXI–XLV (1973).
[22] Sih, G.C., Strain-energy-density factor applied to mixed mode crack problems, *International Journal of Fracture*, 10, pp. 305–321 (1974).
[23] Haigh, B.P., The strain energy function and the elastic limit, *British Association of Advancement of Sciences*, pp. 486–495 (1919).
[24] Nadai, *Plasticity*, McGraw-Hill Book Co. (1931).
[25] Theocaris, P.S., Experimental determination of the core region in mixed-mode fracture, *Proceedings of the First USA–Greece Symposium on "Mixed Mode Crack Propagation'*, Athens, Greece, August 18–22, 1980, Editors G.C. Sih and P.S. Theocaris, Sijthoff and Noordhoff, pp. 21–36 (1981).
[26] Theocaris, P.S., The caustic as a means to define the core region in brittle fracture, *Engineering Fracture Mechanics*, 14, pp. 353–362 (1981).
[27] Sih, G.C. and Macdonald, B., Fracture mechanics applied to engineering problems – strain energy density fracture criterion, *Engineering Fracture Mechanics*, 6, pp. 361–386 (1974).
[28] Gillemot, L.F., Criterion of crack initiation and spreading, *Engineering Fracture Mechanics*, 8, pp. 239–253 (1976).
[29] Kipp, M.E. and Sih, G.C., The strain energy density failure criterion applied to notched elastic solids, *International Journal of Solids and Structures*, 11, pp. 153–173 (1975).

2

The general two-dimensional crack problem

2.1. Introduction

In the present chapter the equations of the strain energy density criterion for the general two-dimensional crack problem will be developed. Using these equations, the critical characteristic quantities of brittle fracture incorporating the critical load for the onset of rapid crack extension and the angle formed by the new branch of the crack with its initial direction are determined. The load for crack initiation is calculated in terms of the critical strain energy density factor S_{cr}, a parameter that accounts for the brittle fracture behavior of a material.

The equations of the two-dimensional crack problem will then be used for the solution of the problem of a single crack in a plate subjected to various loads which generally produce mixed-mode conditions at the crack tips. First the problem of a crack in an infinite plate subjected to a uniform uniaxial stress at infinity at any angle of inclination with respect to the crack axis will be considered. Then we will study the case of a plate of finite width, which is more common in practical applications. For this case the influence of the ratio of the plate width to the crack length on the crack extension angle is determined. This chapter concludes with the case of an infinite plate with a crack subjected to a concentrated force or moment at an arbitrary position with respect to the crack.

2.2. Strain energy density theory for the two-dimensional crack problem

Consider the general problem of a crack in a two-dimensional mixed-mode stress field governed by the values of the opening-mode, k_1, and sliding-mode, k_2, stress intensity factors (Figure 2.1). For this case under generalized plane stress conditions, the singular stress components σ_x, σ_y, τ_{xy} in the vicinity of the crack tip for a system of Cartesian coordinates with the x-axis directed along the crack and the y-axis perpendicular to the crack plane are given by the following equations [1]:

$$\sigma_x = \frac{k_1}{(2r)^{1/2}} \cos\frac{\theta}{2}\left(1 - \sin\frac{\theta}{2}\sin\frac{3\theta}{2}\right) - \frac{k_2}{(2r)^{1/2}} \sin\frac{\theta}{2}\left(2 + \cos\frac{\theta}{2}\cos\frac{3\theta}{2}\right) \quad (2.1a)$$

$$\sigma_y = \frac{k_1}{(2r)^{1/2}} \cos\frac{\theta}{2}\left(1 + \sin\frac{\theta}{2}\sin\frac{3\theta}{2}\right) + \frac{k_2}{(2r)^{1/2}} \sin\frac{\theta}{2}\cos\frac{\theta}{2}\cos\frac{3\theta}{2} \quad (2.1b)$$

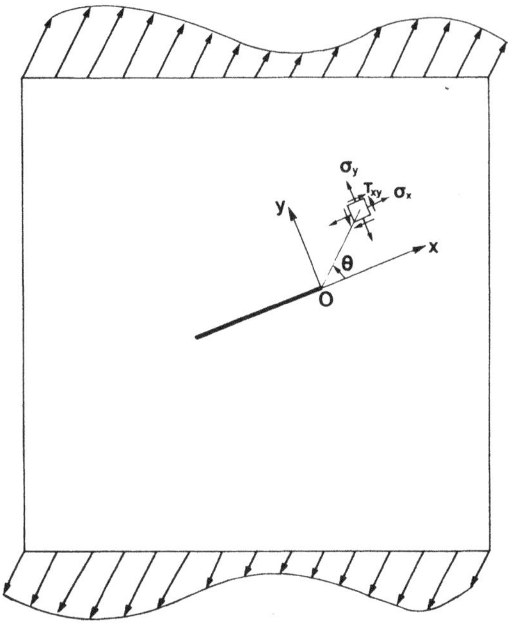

Figure 2.1. A crack in a two-dimensional mixed-mode stress field.

$$\tau_{xy} = \frac{k_1}{(2r)^{1/2}} \cos\frac{\theta}{2} \sin\frac{\theta}{2} \cos\frac{3\theta}{2} + \frac{k_2}{(2r)^{1/2}} \cos\frac{\theta}{2}\left(1 - \sin\frac{\theta}{2} \sin\frac{3\theta}{2}\right) \tag{2.1c}$$

$$\sigma_z = \tau_{xz} = \tau_{yz} = 0 \tag{2.1d}$$

where r, θ represent the polar coordinates of the point considered.

For plane strain conditions σ_z is not equal to zero ($\sigma_z = \nu(\sigma_x + \sigma_y)$), and is given by:

$$\sigma_z = 2\nu \frac{k_1}{(2r)^{1/2}} \cos\frac{\theta}{2} - 2\nu \frac{k_2}{(2r)^{1/2}} \sin\frac{\theta}{2} \tag{2.1e}$$

where ν represents the Poisson's ratio.

Introducing equations (2.1) into (1.3) the following quadratic form for the strain energy per unit volume $\mathrm{d}W/\mathrm{d}V$ is obtained:

$$\frac{\mathrm{d}W}{\mathrm{d}V} = \frac{1}{r}(a_{11}k_1^2 + 2a_{12}k_1k_2 + a_{22}k_2^2) \tag{2.2}$$

where the coefficients a_{ij} $(i, j = 1, 2)$ are given by:

$$16\mu a_{11} = (1 + \cos\theta)(\kappa - \cos\theta) \tag{2.3a}$$

$$16\mu a_{12} = \sin\theta\,[2\cos\theta - (\kappa - 1)] \tag{2.3b}$$

$$16\mu a_{22} = (\kappa + 1)(1 - \cos\theta) + (1 + \cos\theta)(3\cos\theta - 1) \tag{2.3c}$$

with $\kappa = 3 - 4\nu$ or $\kappa = (3-\nu)/(1+\nu)$ for plane strain or plane stress conditions respectively. It is observed that the function $\mathrm{d}W/\mathrm{d}V$ is inversely proportional to the radial distance r measured from the crack tip and it increases as r decreases ($\mathrm{d}W/\mathrm{d}V$ becomes unbounded as $r \to \infty$).

Substituting equation (2.2) into (1.2) we obtain for the strain energy density factor S

$$S = a_{11}k_1^2 + 2a_{12}k_1k_2 + a_{22}k_2^2. \tag{2.4}$$

This factor depends on θ through the coefficients a_{ij} and describes the local energy density on any radial plane intersecting the crack tip.

Equations (1.10) and (1.11) of the strain energy density criterion take the form

$$\frac{\partial S}{\partial \theta} = 0 \tag{2.5a}$$

$$\frac{\partial^2 S}{\partial \theta^2} > 0 \tag{2.5b}$$

and substituting S from equation (2.4) we obtain

$$[2\cos\theta - (\kappa - 1)]\sin\theta\, k_1^2 + 2[2\cos 2\theta - (\kappa-1)\cos\theta]k_1k_2 + (\kappa - 1 - 6\cos\theta)\sin\theta\, k_2^2 = 0 \tag{2.6a}$$

$$[2\cos 2\theta - (\kappa-1)\cos\theta]k_1^2 + 2[(\kappa-1)\sin\theta - 4\sin 2\theta]k_1k_2 + [(\kappa-1)\cos\theta - 6\cos 2\theta]k_2^2 > 0 \tag{2.6b}$$

Relations (2.6a) and (2.6b) represent the general formulae of the strain energy density criterion for the case of a crack in a two-dimensional stress field under mixed-mode loading conditions. For a particular problem under consideration with known values of the k_1, k_2 stress intensity factors introducing these values into equation (2.6a) we obtain the values of the crack extension angle θ_0 as the roots of this equation which satisfy inequality (2.6b). Substituting these roots, θ_0, into equation (2.4) the minimum values $S_{\min}$ of the strain energy density factor are obtained. When $S_{\min}$ is equated to the critical strain energy density factor S_{cr}, which is a material constant, the critical values of the applied loads corresponding to the onset of rapid crack propagation are obtained.

In this section the above procedure will be used for the determination of the angle of initial crack extension and the corresponding critical load for the case of a single crack in a plate under various loading conditions.

2.3. The infinite plate with an inclined crack under uniform uniaxial stress

Consider a central crack of length $2a$ in an infinite plate subjected to a uniform uniaxial stress σ at infinity where the axis of the crack makes an angle β with the direction of stress σ. Mixed-mode conditions predominate in the vicinity of the crack tip, and the values of the k_1, k_2 stress intensity factors are given by [2]:

$$k_1 = \sigma a^{1/2} \sin^2\beta \qquad k_2 = \sigma a^{1/2} \sin\beta \cos\beta. \tag{2.7}$$

Substituting these values into equation (2.4) the following equation is obtained for the strain energy density factor S:

$$S = \sigma^2 a(a_{11} \sin^2\beta + 2a_{12} \sin\beta \cos\beta + a_{22} \cos^2\beta) \sin^2\beta \tag{2.8}$$

where the coefficients a_{ij} are given by equations (2.3).

Equation (2.6a) for the calculation of the angle θ_0 of initial crack extension takes the form:

$$(\kappa - 1) \sin(\theta_0 - 2\beta) - 2 \sin[2(\theta_0 - \beta)] - \sin 2\theta_0 = 0 \qquad \beta \neq 0. \tag{2.9}$$

Before proceeding to the general case of a crack of any inclination with respect to the loading direction, we will consider separately the two common cases of opening-mode and sliding-mode crack extension.

(a) *Opening-mode crack extension*: This case corresponds to the trivial Griffith crack configuration consisting of an infinite body with a central crack of length $2a$ subjected to a uniform uniaxial stress σ at infinity (Figure 2.2a). Because of load symmetry the crack propagates in its own plane. Let us suppose that this is an unknown problem and analyze it through the use of the strain energy density theory. Inserting the values of the stress intensity factors k_1, k_2:

$$k_1 = \sigma a^{1/2} \qquad k_2 = 0 \tag{2.10}$$

into equation (2.4) the following equation is obtained:

$$S = \frac{\sigma^2 a}{16\mu}(1 + \cos\theta)(\kappa - \cos\theta). \tag{2.11}$$

Furthermore, equation (2.6a) which gives the stationary values of S takes the form:

$$[2\cos\theta - (\kappa - 1)] \sin\theta = 0 \tag{2.12}$$

while inequality (2.6b) becomes:

$$2\cos 2\theta - (\kappa - 1)\cos\theta > 0. \tag{2.13}$$

Equation (2.12) is satisfied when $\theta_0 = 0$ or $\theta_0 = \arccos[(\kappa - 1)/2]$. The second root θ_0 does not satisfy inequality (2.13) because for the elastic constant κ is $1 \leqslant \kappa \leqslant 3$, and thus it is disregarded. Hence, the minimum value of S corresponds to an initial crack extension angle $\theta_0 = 0$, which means that the crack extends in its own plane. The plane ($\theta_0 = 0$) corresponds to the direction of maximum potential energy, a position of unstable equilibrium. For $\theta_0 = 0$ equation (2.11) gives the minimum value S_{min} of S

$$S_{min} = \frac{(\kappa - 1)\sigma^2 a}{8\mu}. \tag{2.14}$$

Equating S_{min} with S_{cr}, which is a material constant, we obtain the following expression for the critical stress σ_{cr} corresponding to the onset of crack extension

$$\sigma_{cr} a^{1/2} = \left(\frac{8\mu S_{cr}}{\kappa - 1}\right)^{1/2}. \tag{2.15}$$

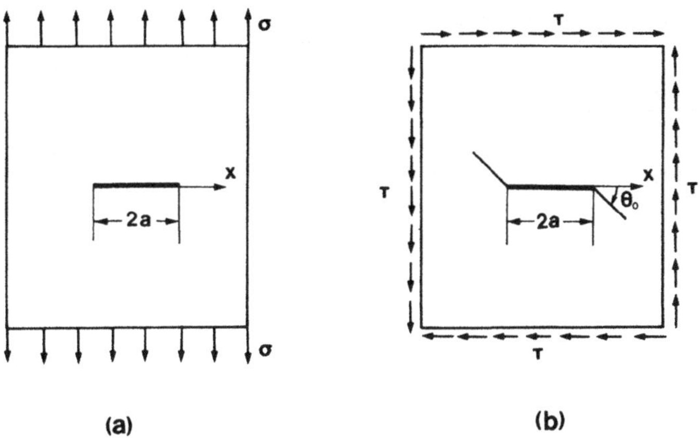

Figure 2.2. Opening-mode (a) and sliding-mode (b) crack extension.

The quantities entered in the right hand side of equation (2.15) are material constants. Even though the derivation of equation (2.15) is based on an entirely different premise, it is similar in form to equation (1.1) obtained by Griffith. Thus, for cracks propagating in a self-similar manner, S_{cr} can be related to γ in Griffith theory and to k_{1c} or G_{1c} in the Griffith–Irwin theory. From relations (2.15) and (1.1) and for plane strain conditions the following equation is obtained:

$$\sigma_{cr}^2 a = \left[\frac{2E}{(1+\nu)(1-2\nu)}\right] S_{cr} = \left[\frac{2E}{\pi(1-\nu^2)}\right] \gamma = k_{1c}^2. \qquad (2.16)$$

(b) *Sliding-mode crack extension*: This case corresponds to an infinite body containing a central crack of length $2a$ and subjected to a uniform shear stress τ at infinity (Figure 2.2b). The k_1, k_2 stress intensity factors are given by:

$$k_1 = 0 \qquad k_2 = \tau a^{1/2} \qquad (2.17)$$

and equation (2.4) gives:

$$S = \frac{\tau^2 a}{16\mu} [(\kappa + 1)(1 - \cos\theta) + (1 + \cos\theta)(3\cos\theta - 1)]. \qquad (2.18)$$

Furthermore, equation (2.6a) which gives the stationary values of S takes the form:

$$(\kappa - 1 - 6\cos\theta)\sin\theta = 0 \qquad (2.19)$$

while inequality (2.6b) becomes:

$$(\kappa - 1)\cos\theta - 6\cos 2\theta > 0. \qquad (2.20)$$

Of the two roots of equation (2.19) $\theta_0 = 0$ and $\theta_0 = \arccos[(\kappa - 1)/6]$ only the second root satisfies inequality (2.20). It is observed that θ_0 is a function of the Poisson's ratio ν. Table 2.1 shows the values of the predicted fracture angle θ_0 for ν ranging from 0 to 0.5 for plane strain conditions ($\kappa = 3 - 4\nu$).

Table 2.1. Fracture angle $-\theta_0$ under pure shear and plane strain conditions.

ν	0	0.1	0.2	0.3	0.4	0.5
$-\theta_0$	70.5°	74.5°	78.5°	82.3°	86.2°	90.0°

Introducing the value of the angle of initial crack extension $\theta_0 = \arccos[(\kappa - 1)/6]$ into equation (2.18) and equating the resulting value of $S_{\min}$ to S_{cr} the following expression for the critical shear stress τ_{cr} is derived:

$$\tau_{cr} a^{1/2} = \left(\frac{192 \mu S_{cr}}{-\kappa^2 + 14\kappa - 1} \right)^{1/2}. \tag{2.21}$$

It should be observed that equation (2.21) is similar in form to equation (2.15) for the crack under symmetric loads.

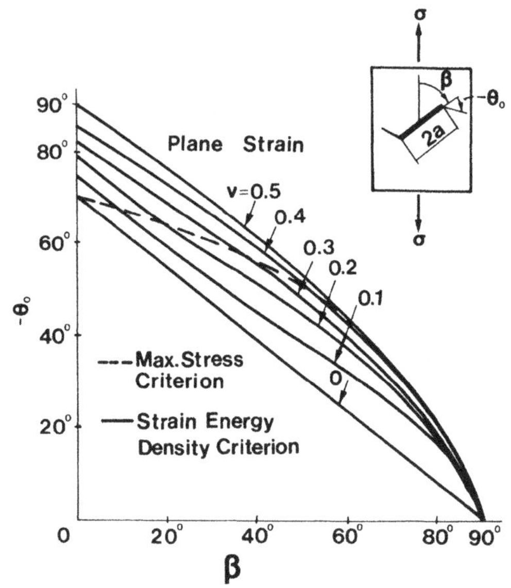

Figure 2.3. Variation of the crack extension angle $-\theta_0$ versus β under plane strain conditions for tensile applied loads.

(c) *The inclined crack; tensile loads*: Solution of equation (2.9) with the crack inclination angle β ranging from 0 to 90° and various values of the material constant κ provides the stationary value of the strain energy density factor. From the resulting values of the angle θ, those which satisfy inequality (2.6b) provide the angles of initial crack extension θ_0. Figure 2.3 displays the variation of $-\theta_0$ versus the crack angle β for $\nu = 0$, 0.1, 0.2, 0.3, 0.4 and 0.5 under plane strain conditions. Results for plane stress conditions can be obtained by replacing ν with $\nu/(1 + \nu)$. In the same figure the dotted curve represents the results obtained by the maximum stress criterion [3]. It is observed that these results agree with those based on equation (2.9) for large values of β and represent a lower bound for small values of β. In general it can be taken as an

Table 2.2. Measured and calculated values of the fracture angle.

β		30°	40°	50°	60°	70°	80°
	1	−64°	−55.5°	−50°	−40°	−29°	−17°
θ_0	2	−60°	−52°	−50°	−43.5°	−30.5°	−18°
@ right	3	−63°	−57°	−53°	−44.5°	—	−15.5°
	4	—	−57°	−52°	−43.5°	—	—
	1	−65°	−58°	−50.5°	−44°	−31.5°	−18.5°
θ_0	2	—	−53°	−52°	−40°	−31°	−17.5°
@ left	3	−60°	−55°	−51.5°	−46°	−31.5°	−17°
	4	—	−57°	−50°	−43°	—	—
$(\theta_0)_{avg}$		−62.4°	−55.6°	−51.1°	−43.1°	−30.7°	−17.3°
θ_0^*		−60.2°	−55.7°	−50.2°	−43.2°	−33.2°	−19.3°
θ_0^+		−63.5°	−56.7°	−49.5°	−41.5°	−31.8°	−18.5°

θ_0^* obtained from the maximum stress criterion.
θ_0^+ obtained from equation (2.9) of the strain energy density criterion.

average curve. It is worth noting that the crack extension angle θ_0 is always negative for uniaxial tensile loads, as is depicted by Figure 2.3.

The results of Figure 2.3 were checked by a series of experiments performed on plexiglas plates with a central crack. The dimensions of the specimens were 9″ x 18″ x 3/16″ and the crack had a length $2a$ of approximately 2″, forming angles with the direction of the applied tensile stress β ranging from 30° to 80° in increments of 10°. The angles of initial crack extension θ_0 in both tips (right and left) were measured. The experimental data for four sets of tests are given in Table 2.2 with $(\theta_0)_{avg}$ being the average fracture angle of all the measured values. The last two rows give the theoretical results according to the maximum stress criterion and the strain energy density criterion with $\nu = 1/3$. It can be observed that the experimental results are in good agreement with the theoretical predictions.

Introducing the values of the crack extension angle θ_0 as they are displayed in Figure 2.3 into equation (2.8) the minimum values of the strain energy density factor S are obtained. Equating these values of S with the critical strain energy density factor S_{cr}, which is a material constant, the values of the critical tensile stress σ_{cr} for crack propagation are obtained. The variation of the quantity $16\mu S_{cr}/\sigma_{cr}^2 a$ versus the crack inclination angle β for $\nu = 0$, 0.1, 0.2, 0.3, 0.333 and 0.4 is shown in Figure 2.4 when plane strain conditions prevail in the plate. It is observed that the quantity $16\mu S_{cr}/\sigma_{cr}^2 a$ increases with the crack angle β, reaching a maximum for opening-mode crack extension. Furthermore, $16\mu S_{cr}/\sigma_{cr}^2 a$ increases as the Poisson's ratio ν of the plate decreases. Since S_{cr} is a material constant, the above statements imply that the quantity $\sigma_{cr}^2 a$ decreases as the crack angle β increases, while quite the contrary happens with respect to the Poisson's ratio ν. Thus, the lowest value of the applied stress σ_{cr} which will initiate crack propagation occurs at $\beta = \pi/2$ for a material with a low Poisson's ratio.

These theoretical results were compared with experiments made on DTD 5050–5½% Zn aluminum alloy in [4]. All the tests were carried out on specimens with inclined cracks. The measured values of $\sigma_{cr}\sqrt{a}$ for various values of the crack inclination

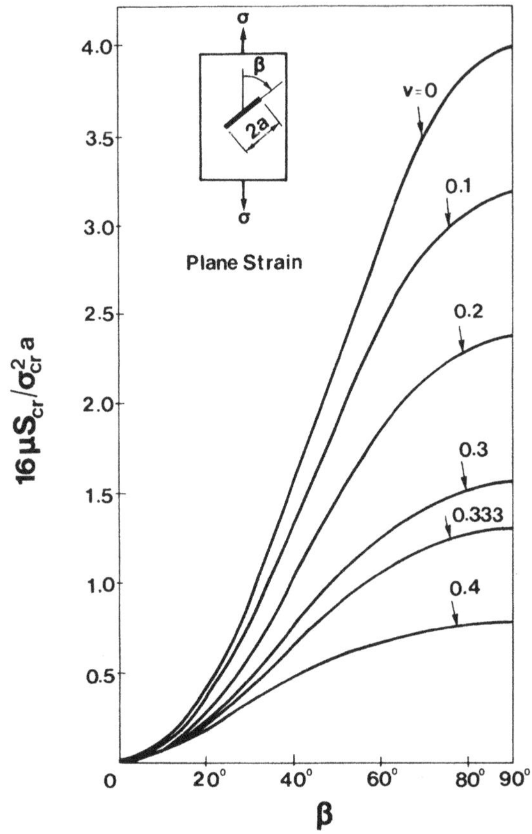

Figure 2.4. Variation of the quantity $16\mu S_{cr}/\sigma_{cr}^2 a$ versus the crack angle β under plane strain conditions for tensile applied loads.

angle β are shown in Figure 2.5. The theoretical curves (solid and dotted lines) were obtained for values of S_{cr} corresponding to the aluminum alloy with k_{1c} equal to 28.2 ksi in$^{1/2}$ and 29.2 ksi in$^{1/2}$ respectively (the values of S_{cr} were calculated from k_{1c} through equation (2.16)). It is observed that the experimental results are in good agreement with the theoretical predictions.

(d) *The inclined crack; compressive loads*: In addition to the negative roots of equation (2.9) which correspond to uniaxial tensile loads, there exists another set of solutions for positive angles θ_0. Physically the positive crack extension angles correspond to the case when the inclined crack is under uniaxial compression. Since S depends on σ^2 equations (2.8) and (2.9) contain both the solutions of uniaxial tension $+\sigma$ and compression $-\sigma$. At this point it should be emphasized that for the case of compression it is assumed that no overlapping between the crack lips takes place.

Figure 2.6 presents the variation of the positive crack extension angle θ_0 for compressive applied loads versus the crack inclination angle β for various values of the Poisson's ratio ν. It is observed that contrary to tensile loading where the crack tends

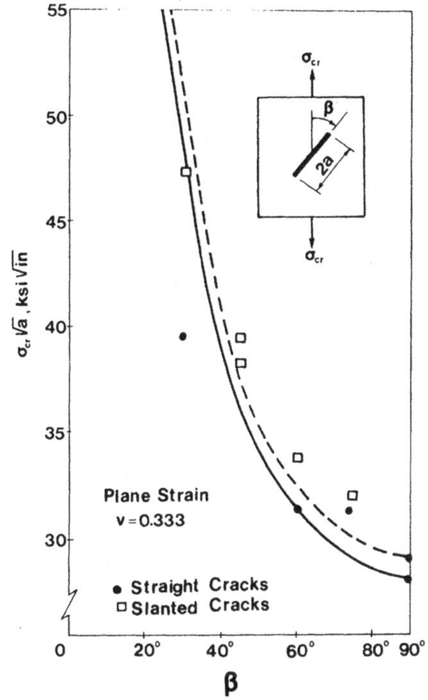

Figure 2.5. Variation of the critical tensile stress $\sigma_{cr}\sqrt{a}$ for crack propagation versus the crack inclination angle β for the aluminum alloy with $k_{1c} = 28.2$ ksi in$^{1/2}$ (solid line) and $k_{1c} = 29.2$ ksi in$^{1/2}$ (dotted line).

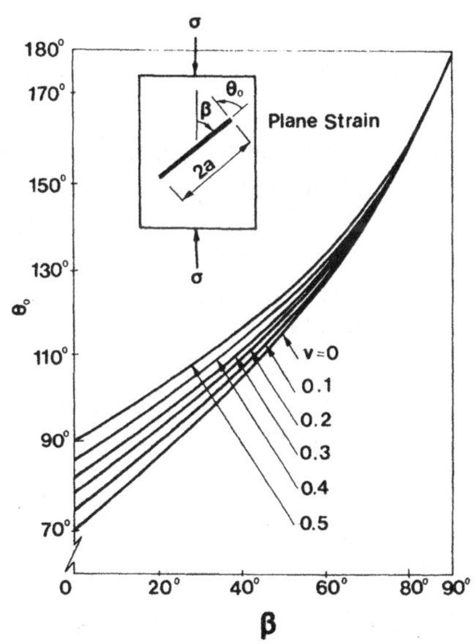

Figure 2.6. Variation of the crack extension angle θ_0 versus β under plane strain conditions for compressive applied loads.

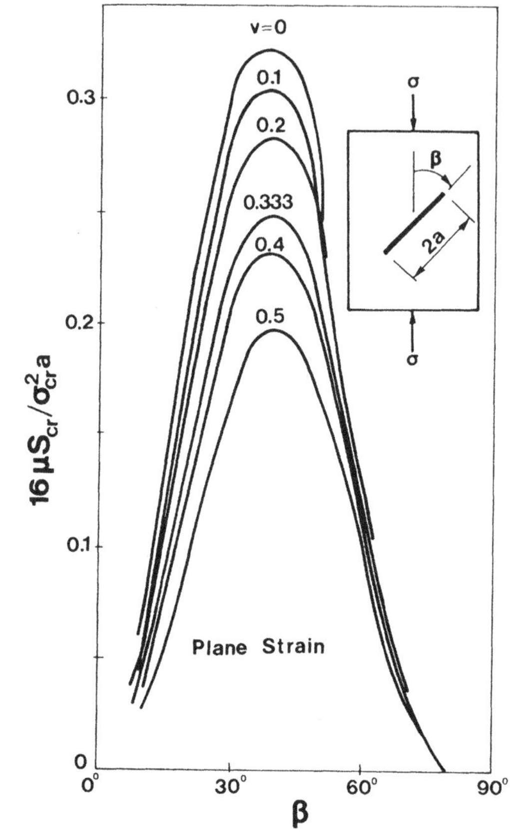

Figure 2.7. Variation of the quantity $16\mu S_{cr}/\sigma_{cr}^2 a$ versus the crack angle β under plane strain conditions for compressive applied loads.

to become horizontal, the crack path under uniaxial compression extends towards the direction of loading. Such a phenomenon has indeed been observed by Hoek and Bieniawski [5] who made tests on a number of glass plates with inclined cracks under uniaxial compression. Unfortunately, they did not report the angle of initial crack extension and therefore a comparison of their experimental results with the theoretical results of Figure 2.6 is not possible.

Following the same procedure as used in the case of tensile loading, the stationary values of the strain energy density factor are obtained. Figure 2.7 presents the variation of the quantity $16\mu S_{cr}/\sigma_{cr}^2 a$ versus the crack inclination angle β for plane strain conditions. From Figure 2.7 we observe that the quantity $16\mu S_{cr}/\sigma_{cr}^2 a$ reaches a maximum in the interval $0 < \beta < 90°$ which depends on the value of the Poisson's ratio ν. It is also observed that the critical stress σ_{cr} increases and tends to infinity as the crack becomes parallel ($\beta = 0$) or perpendicular ($\beta = 90°$) to the direction of loading. This result corresponds to the physical observation that a crack in a plate parallel or perpendicular to the direction of a compressive applied stress has no influence on the fracture behavior of the plate.

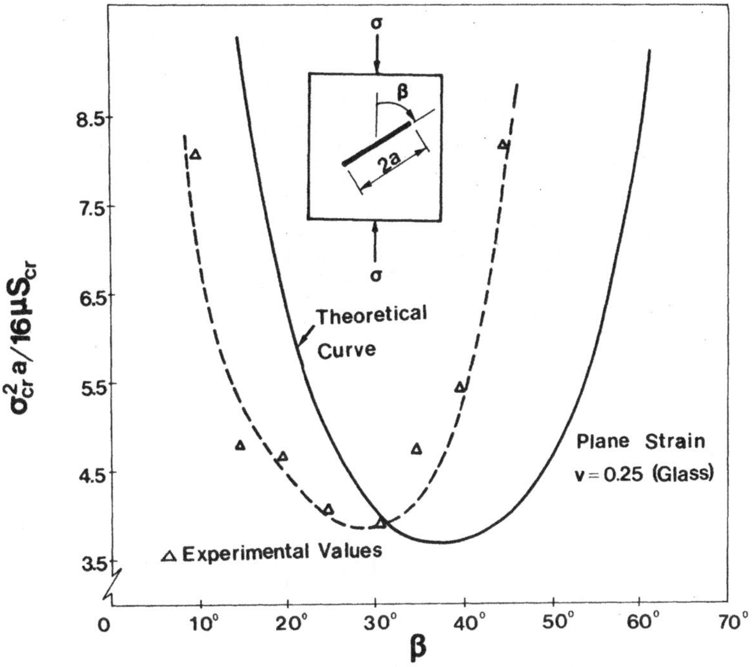

Figure 2.8. Variation of the normalized critical compressive strength $\sigma_{cr}^2 a/16\mu S_{cr}$ versus crack angle β for a precracked glass plate (dotted lines) and the corresponding theoretical predictions based on the strain energy density criterion.

The theoretically predicted results on the critical fracture stress σ_{cr} were compared with experimental data on precracked glass plates of 6″ x 6″ [5]. Figure 2.8 presents the experimental results together with the corresponding theoretical predictions. For the glass plate with $\nu = 0.25$ the theoretical curve of Figure 2.8 predicts an angle of $\beta_0 \simeq 37°$ at which the applied stress to initiate fracture is minimum, while the corresponding experimental curve indicates a critical angle $\beta_0 \simeq 31°$. For a crack loaded under remote compression the crack surfaces tend to come into contact and rub against one another. Thus, the significant point to be made here may not be in the quantitative agreement of theory and experiment but in the trend of the failure stress variations with crack angle for fracture under compression which has been predicted by the strain energy density theory. This theory also shows that the apparent compressive strength of brittle materials can be many times greater than its tensile strength, depending upon geometry and loading conditions.

For more details on the results of this section the interested reader is referred to references [6–10].

2.4. Finite width effects on the crack extension angle

While in the previous section the brittle fracture characteristic quantities of an infinite plate subjected to a remote uniform uniaxial stress were studied, it is the objective

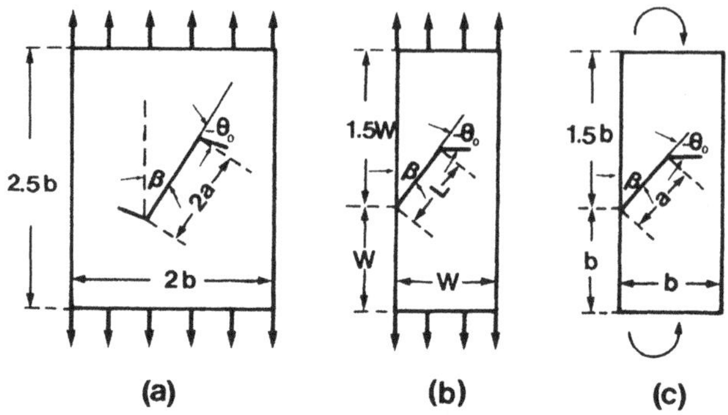

Figure 2.9. Geometry of the centre- (a) and edge-cracked plates in tension (b) and bending (c).

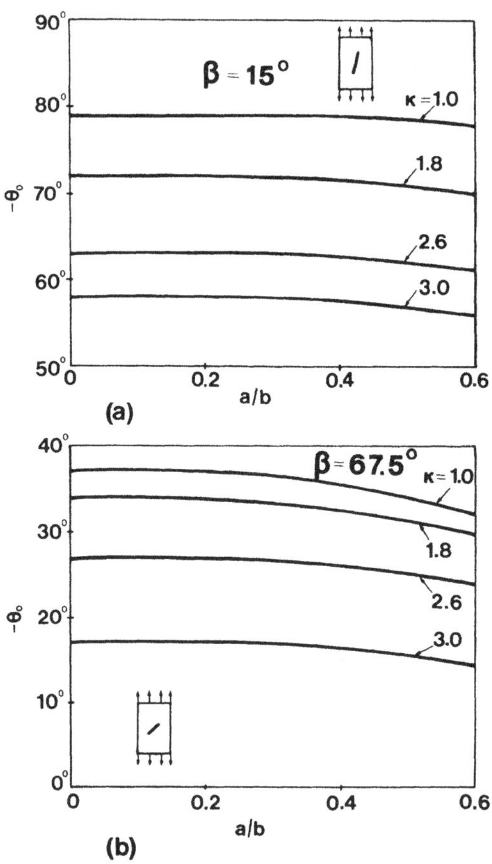

Figure 2.10. Variation of the crack extension angle $-\theta_0$ versus the ratio a/b of crack length to plate width for a center-cracked specimen with $\beta = 15°$ (a) and $\beta = 67.5°$ (b) under tensile applied loads.

of the present section to analyze the influence of the plate width on the crack extension angle. Finite width plates are more commonly encountered in practical applications than plates with large dimensions whose boundaries do not generally affect the fracture quantities at the crack tip, and which therefore can be considered as infinitely large. The finite width, center- and edge-cracked plates under tension and bending studied in this section are shown in Figure 2.9. The values of k_1 and k_2 stress intensity factors for the center-cracked plate in tension were determined by Wilson [11]. These values for the edge-cracked plate in tension and in bending were given by Bowie [12] and Wilson [13] respectively. For each pair of values k_1 and k_2 and for the values of the material constant κ equal to 1.0, 1.8. 2.6 and 3.0 (corresponding to the values of the Poisson's ratio ν equal to 0.5, 0.3, 0.1 and 0 respectively under plane strain conditions) equation (2.6a) was solved in a computer. Four roots θ_0 are generally obtained for each case. Two of them are rejected because they do not satisfy the inequality of relation (2.6b). Of the remaining crack extension angles one is negative, corresponding to tensile loads and the other is positive, for compressive loads.

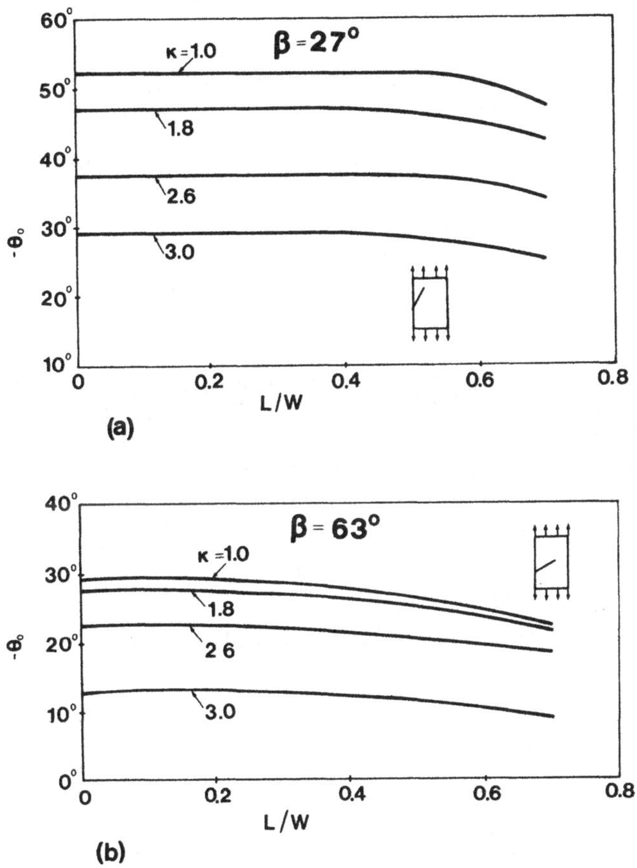

Figure 2.11. As in Figure 2.10 for an edge-cracked plate in tension with $\beta = 27°$ (a) and $\beta = 63°$ (b).

Figure 2.10 presents the variation of the crack extension angle $-\theta_0$ versus the ratio a/b of crack length to plate width for the center-cracked plate of Figure 2.9a subjected to tensile loads, and for the values of the crack inclination angle relative to the direction of the applied load equal to 15° and 67.5°. The values of the angle $-\theta_0$ for the edge-cracked plate of Figure 2.9b subjected to tensile loads for $\beta = 27°$ and 63° are displayed in Figure 2.11. Finally, Figure 2.12 gives the variation of the angle $-\theta_0$ versus a/b for the edge-cracked plate of Figure 2.9c loaded in bending for $\beta = 45°$ and 67.5°.

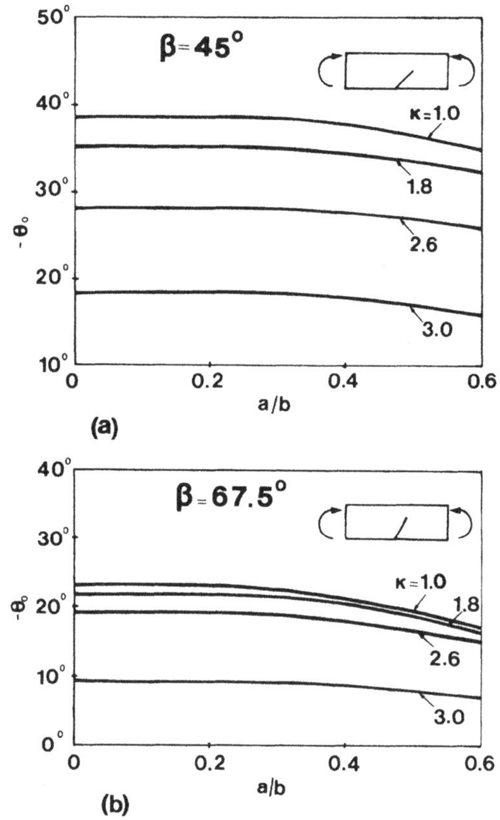

Figure 2.12. As in Figure 2.10 for an edge-cracked plate in bending with $\beta = 45°$ (a) and $\beta = 67.5°$ (b).

From the above figures the dependence of the crack extension angle on the plate geometry can be easily disclosed. Thus, the following conclusions can be drawn:

1. The crack extension angle $-\theta_0$ decreases as the crack angle increases.
2. The dependence of $-\theta_0$ on crack length is insignificant for the values of the ratio a/b of the crack length to the specimen's width lying in the interval $0 < a/b < 0.2$.
3. The decrease of $-\theta_0$ with the increase of the ratio a/b is more rapid as the crack inclination angle β with respect to the direction of loading increases.

Furthermore, the decrease of $-\theta_0$ is greater for the edge- than for the center-cracked specimens.

4. The relationship between $-\theta_0$ and the crack inclination angle for the edge- and the center-cracked specimens is almost similar. However, the values of $-\theta_0$ for the edge-cracked specimens are much smaller than for the center-cracked specimens.
5. The values of $-\theta_0$ for the edge-cracked specimen in bending are closer to the values of $-\theta_0$ for the same specimen in tension than to those of the center-cracked specimen in tension. This indicates that the geometry of the specimen influences more the results than the form of loading.

For more details on the dependence of the crack extension angle on plate geometry the interested reader is referred to references [14, 15].

2.5. The cracked plate subjected to a concentrated force or moment

In the present section the critical load for crack propagation and the corresponding crack extension angle of an infinite plate with a single crack subjected to a concentrated force or moment at an arbitrary point are determined. The cases of concentrated force parallel or perpendicular to the crack axis and of the localized moment are considered individually below.

(a) *The cracked plate subjected to a concentrated force parallel to the crack axis*: Consider an infinite elastic plate with a crack of length $2a$ subjected to a concentrated force Q per unit thickness, which is parallel to the crack axis and acting at an arbitrary point K (Figure 2.13). The plate is referred to a system of Cartesian coordinates OxY with its origin at the middle of the crack and the x-axis coinciding with the crack line. Let x, y designate the coordinates of point K.

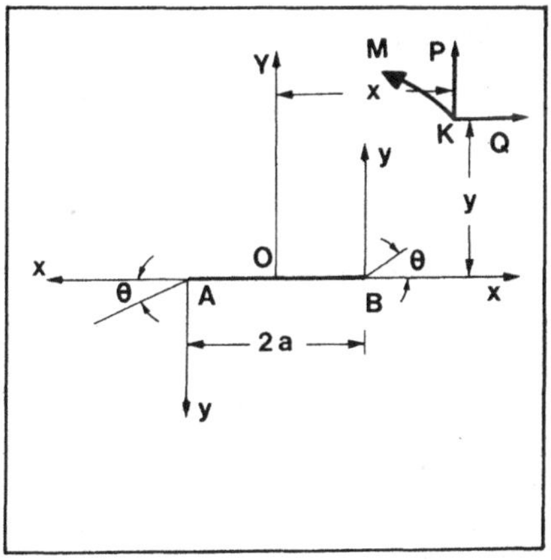

Figure 2.13. Geometry of a cracked plate subjected to a concentrated force or moment at an arbitrary point.

This case was studied by Erdogan [16] and Sih [17] who found the stress state in the neighborhood of the crack tip. The values of the opening-mode k_{1B} and sliding-mode k_{2B} stress intensity factors for the tip B as they were reported by Rooke and Cartwright [18] are expressed by the following relations:

$$k_{1B}(a;x,y) = \left(G_1 + \frac{1}{\kappa - 1} H_1\right)\left(\frac{\kappa - 1}{\kappa + 1}\right) k_0$$
$$k_{2B}(a;x,y) = \left(G_2 + \frac{1}{\kappa + 1} H_2\right) k_0 \tag{2.22a}$$

where:

$$G_1 = 1 - \text{Re}\,[G(z)] \qquad H_1 = \text{Re}\,[H(z)]$$
$$G_2 = -\text{Im}\,[G(z)] \qquad H_2 = -\text{Im}\,[H(z)] \tag{2.22b}$$

with

$$G(z) = (a + z)(z^2 - a^2)^{-1/2}$$
$$H(z) = a(\bar{z} - z)(\bar{z} - a)^{-1}(z^2 - a^2)^{-1/2} \tag{2.22c}$$

and:

$$k_0 = \frac{Q}{2\pi a^{1/2}}. \tag{2.22d}$$

In the above relations z is the complex variable associated with the system OxY ($z = x + iy$) and Re and Im denote real and imaginary parts respectively.

For the special case when the load Q lies on the crack ($y = 0$, $-a \leqslant x \leqslant a$) the values of k_{1B} and k_{2B} as they were determined by Sih et al. [2] are given by:

$$k_{1B} = \left(\frac{\kappa - 1}{\kappa + 1}\right) k_0$$
$$k_{2B} = (1 + x/a)^{-1/2}(1 - x/a)^{-1/2} k_0. \tag{2.23}$$

Due to the symmetry relations the values of the stress intensity factors k_{1A} and k_{2A} for the other tip A are given by:

$$k_{1A}(-a;x,y) = -k_{1B}(a;-x,y)$$
$$k_{2A}(-a,x,y) = k_{2B}(a;-x,y). \tag{2.24}$$

When we introduce the values of the stress intensity factors k_{1B} and k_{2B} for the tip B from equations (2.22) and (2.23) into equation (2.6a) we obtain the values of the angle θ which the fracture path forms with the crack axis for each particular location (x, y) of the point K at which the load Q is applied and for each value of the material constant κ (Figure 2.13). Four roots of this equation are generally found. Of these roots, two are rejected because they do not satisfy the inequality of relation (2.6b). One of the remaining two roots corresponds to the case when the concentrated load is directed to the right (as in Figure 2.13), while the other applies when it is directed to the left. Only the case when the concentrated load is directed to the right was considered. Introducing the values of the crack extension angle θ_{0B} into equation (2.4) the minimum values $S_{\min}$ of the strain energy density factor S normalized to the critical load Q_{crB} required for crack extension are determined. Equating $S_{\min}$ to S_{cr}, which is a material constant, the values of the load Q_{crB} for crack extension are determined.

The same procedure is then used to determine the crack extension angle θ_{0A} and the corresponding load Q_{crA} for the crack tip A. For this reason, the values of the stress intensity factors k_{1A} and k_{2A} given by equation (2.24) are used.

Thus, for a given position K of the point of application of the concentrated load Q we obtain for each crack tip A and B a definite value of the critical load $Q_{crA,B}$ required for crack propagation. It is evident that the smaller Q_{cr} of these two values $Q_{crA,B}$ represents the critical failure load of the cracked plate. Crack propagation initiates from the tip which corresponds to Q_{cr}.

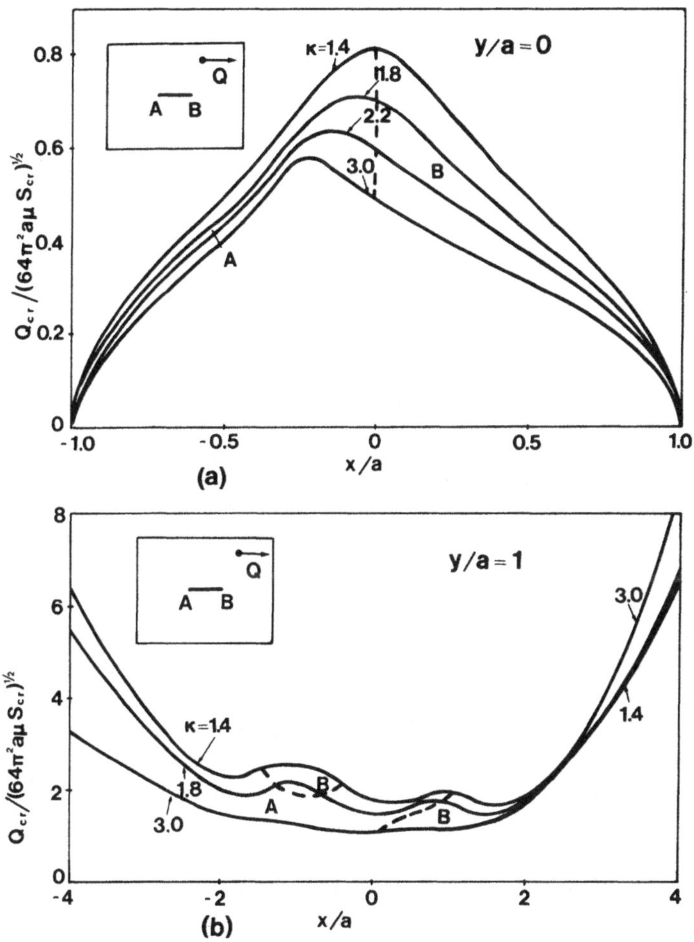

Figure 2.14. Variation of the quantity $Q_{cr}/(64\pi^2 a\mu S_{cr})^{1/2}$ versus x/a for a cracked plate subjected to a concentrated load Q parallel to the crack axis for $y/a = 0$ (a) and $y/a = 1$ (b).

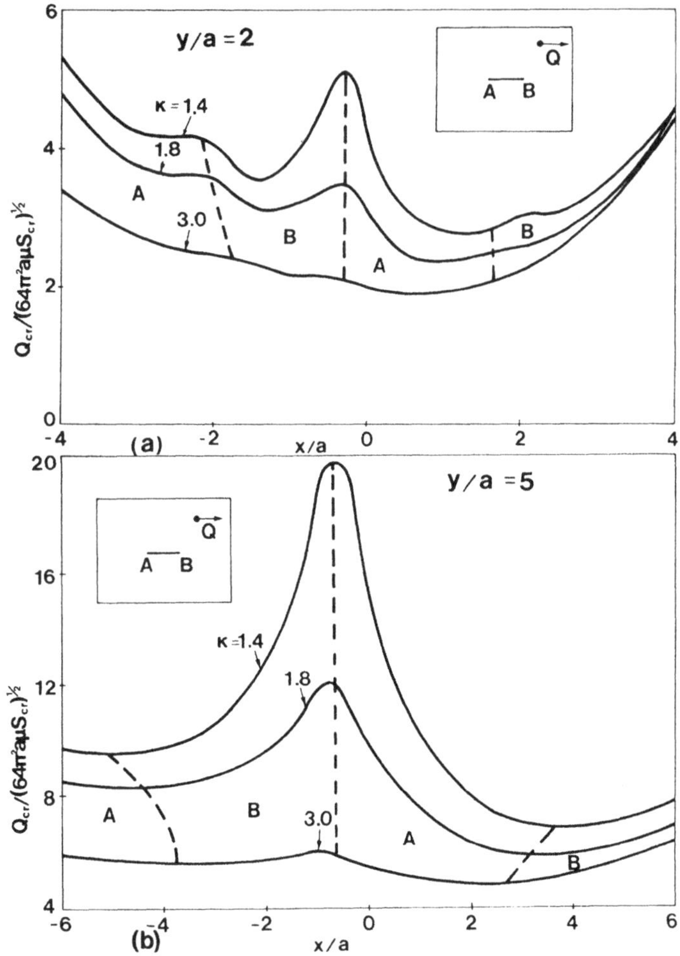

Figure 2.15. As in Figure 2.14 for $y/a = 2$ (a) and $y/a = 5$ (b).

Figures 2.14 and 2.15 present the variation of the quantity $Q_{cr}/(64\pi^2 a\mu S_{cr})^{1/2}$ versus the distance x/a along the crack axis for various values of the perpendicular distance y/a of the point of application of the concentrated load and the material constant κ. The values of the crack extension angle $-\theta_0$ for the cases of Figures 2.14 and 2.15 are displayed in Figures 2.16 and 2.17. In all these figures letters A and B indicate whether crack extension starts from tip A or B. The region in which crack extension begins at crack tip A is separated from the region in which it begins at tip B by dotted lines.

Figure 2.14a shows that the critical load Q_{cr} becomes zero when the concentrated load acts at the crack tips, and takes its maximum value near the middle of the crack. Furthermore Q_{cr} increases for the same value of x/a when the perpendicular distance y/a increases. From Figures 2.14 to 2.17 the dependence of the critical load Q_{cr}, the crack extension angle $-\theta_0$ and the particular crack tip A or B from which crack

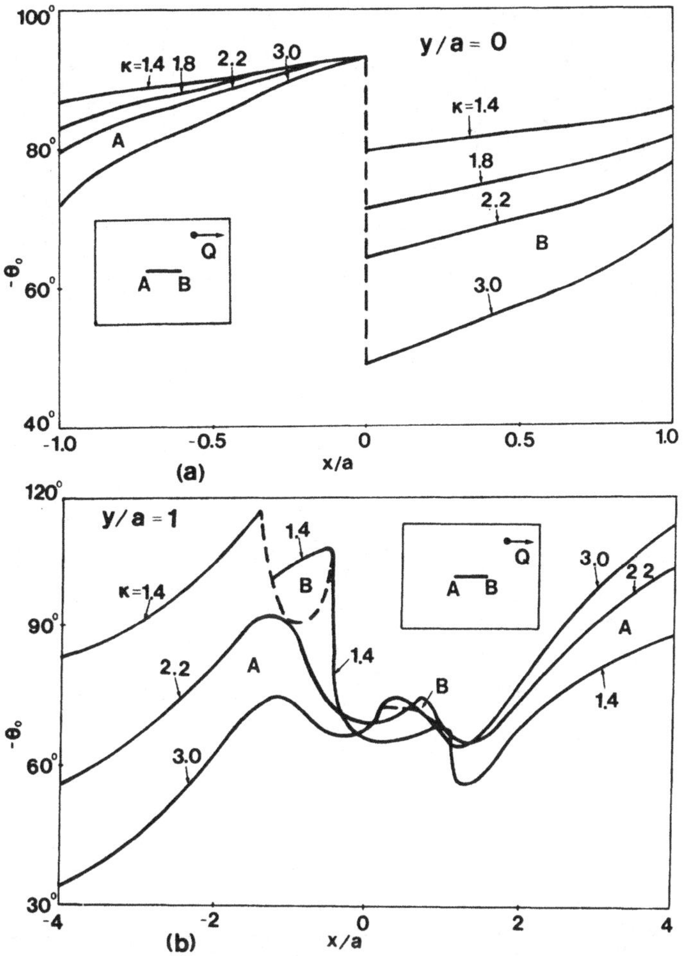

Figure 2.16. Values of the crack extension angle $-\theta_0$ for the case of Figure 2.14.

propagation takes place on the geometrical location of the concentration load and the material constant κ can easily be established.

(b) *The cracked plate subjected to a concentrated force perpendicular to the crack axis*: For the case of Figure 2.13 when the concentrated load P acting at the point K is perpendicular to the crack axis, the values of the stress intensity factors k_{1B} and k_{2B} for the tip B are expressed by the following equations [16–18]:

$$k_{1B}(a;x,y) = \left(G_2 - \frac{1}{\kappa+1} H_2\right) k_0$$
$$k_{2B}(a;x,y) = -\left(G_1 - \frac{1}{\kappa-1} H_1\right)\left(\frac{\kappa-1}{\kappa+1}\right) k_0 \qquad (2.25a)$$

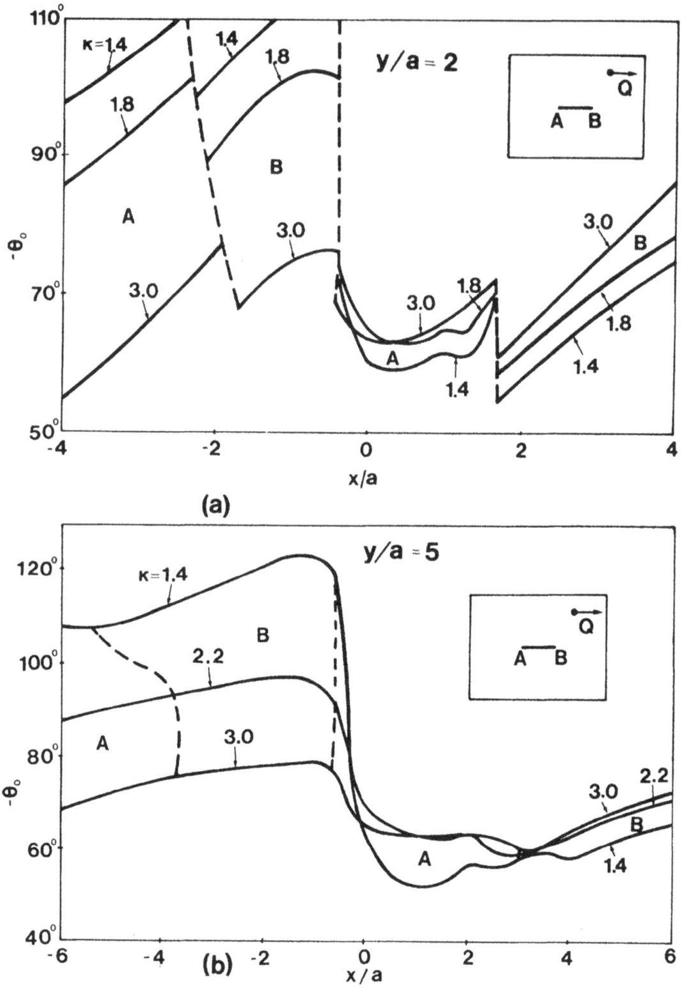

Figure 2.17. Values of the crack extension angle $-\theta_0$ for the case of Figure 2.15.

with:

$$k_0 = \frac{P}{2\pi a^{1/2}} \tag{2.25b}$$

where the values of the functions G_1, G_2, H_1, H_2 are again given by equations (2.22b) and (2.22c).

For the special case when the load P lies on the crack ($y = 0$, $-a \leqslant x \leqslant a$) the values of k_{1B} and k_{2B} are given by [2]:

$$\begin{aligned} k_{1B} &= (1 + x/a)^{1/2}(1 - x/a)^{-1/2} k_0 \\ k_{2B} &= -\left(\frac{\kappa - 1}{\kappa + 1}\right) k_0 . \end{aligned} \tag{2.26}$$

Due to the symmetry relations the values k_{1A} and k_{2A} of the stress intensity factor for the crack tip A are:

$$\begin{aligned} k_{1A}(-a;x,y) &= k_{1B}(a;-x,y) \\ k_{2A}(-a;x,y) &= -k_{2B}(a;-x,y). \end{aligned} \tag{2.27}$$

Following the same procedure as used in the case of concentrated load parallel to the crack axis the critical load for crack extension and the corresponding angle are determined. Two cases are considered: First, the case when the concentrated load is directed in the opposite direction of the crack, and second, when directed toward the crack. Again the results are presented in the form of curves showing the variation of the above quantities for movement of the load on lines parallel to the crack axis. Since all these curves are symmetrical with respect to the axis of symmetry OY of the cracked plate, only the sections corresponding to positive values of the x-coordinate are shown.

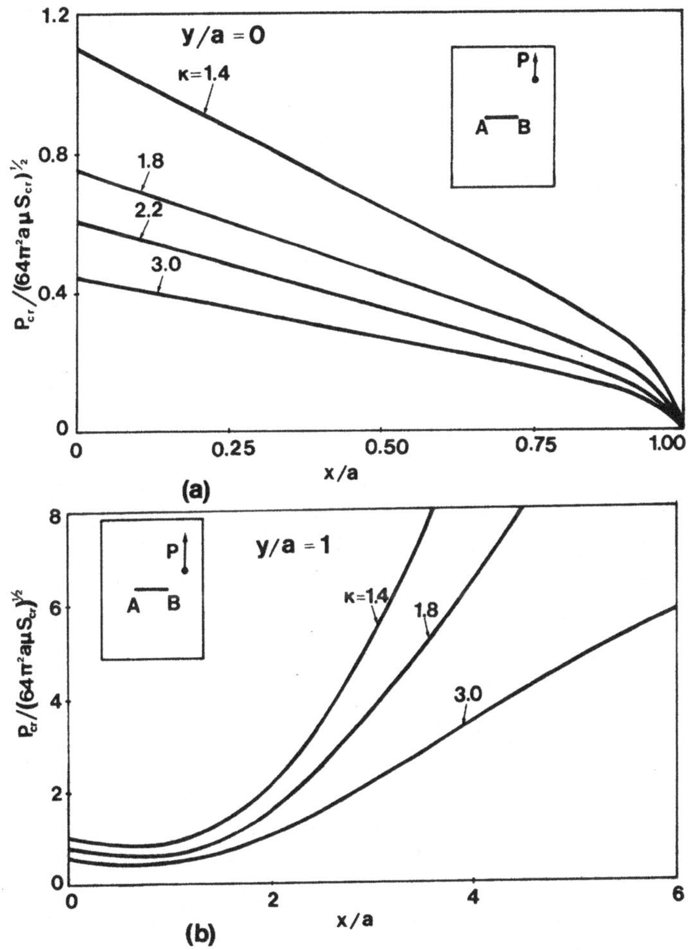

Figure 2.18. Variation of the quantity $P_{cr}/(64\pi^2 a\mu S_{cr})^{1/2}$ versus x/a for a cracked plate subjected to a concentrated load perpendicular to the crack axis and directed in the opposite direction of the crack for $y/a = 0$ (a) and 1 (b).

Figure 2.18 presents the variation of the quantity $P_{cr}/(64\pi^2 a\mu S_{cr})^{1/2}$ versus x/a for $y/a = 0$ and 1.0, and various values of κ for the case when the concentrated load is directed in the opposite direction of the crack. From Figure 2.18a we observe that when the load moves on the crack, its critical value is equal to zero when it lies at the crack tips. It reaches its maximum value when it lies at the middle of the crack in a manner analogous to the case of the concentrated load parallel to the crack. Furthermore from Figure 2.18b we conclude that the minimum value of P_{cr} is found when the load P approaches the crack tip. P_{cr} then increases rapidly as the load moves away from this point. In both cases, showing in Figure 2.18, crack propagation takes place from the crack tip B, which is closer to the load P than the other crack tip A. The variation of the corresponding crack extension angles θ_1 with x/a for the cases of Figure 2.18 is shown in Figure 2.19.

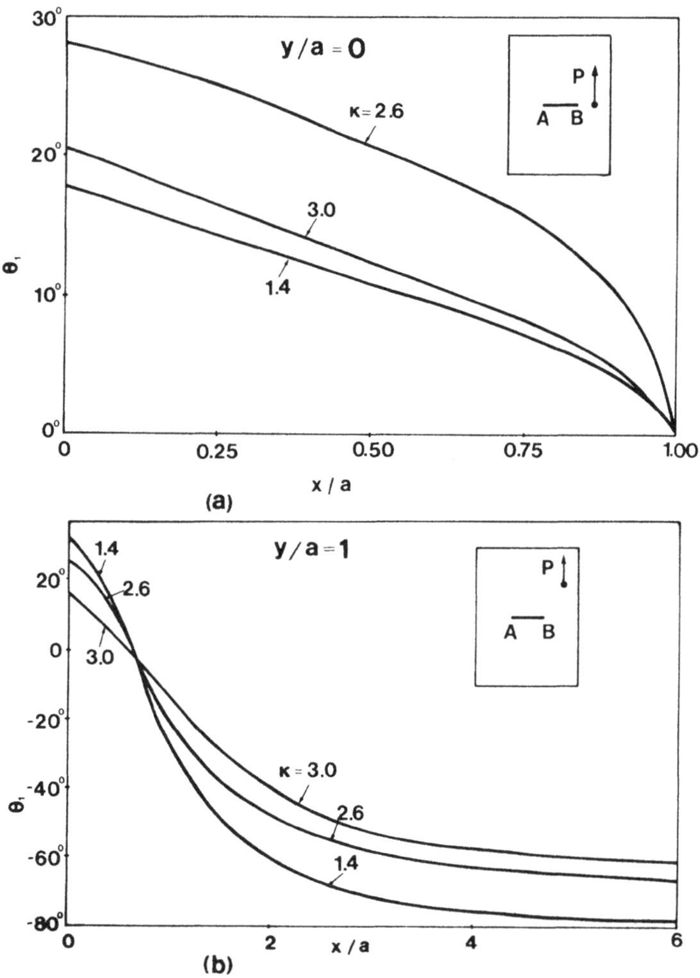

Figure 2.19. Values of the crack extension angle θ_1 for the case of Figure 2.18.

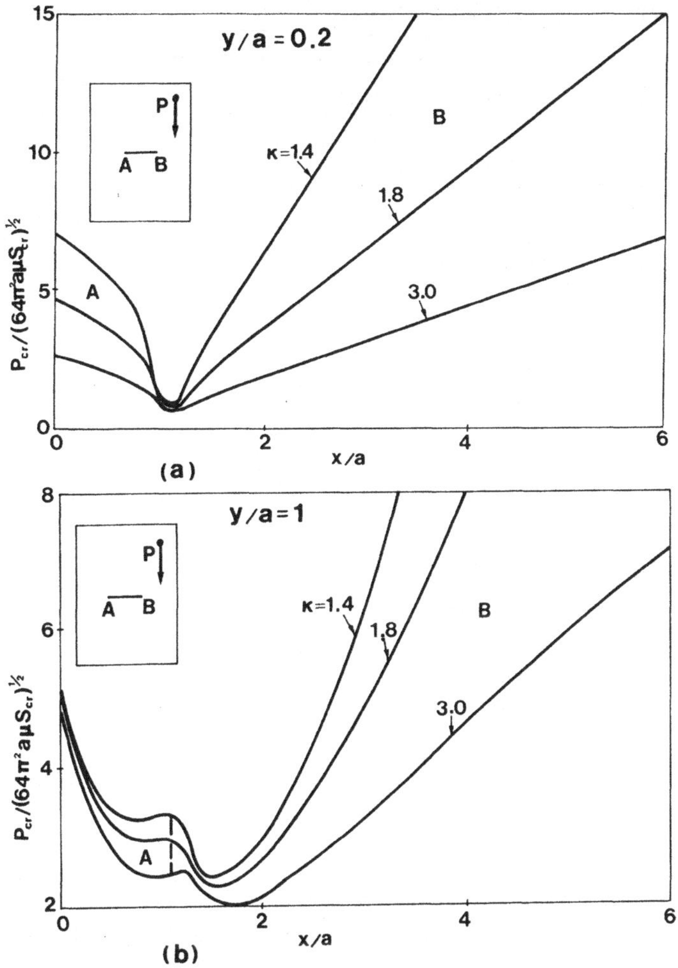

Figure 2.20. As in Figure 2.18 with the concentrated load directed toward the crack for $y/a = 0.2$ (a) and 1.0 (b).

Figure 2.20 presents the variation of the quantity $P_{cr}/(64\pi^2 a\mu S_{cr})^{1/2}$ when the load P is directed toward the crack for $y/a = 0.2$ and 1.0. We observe that for the case of Figure 2.20 crack extension does not always take place from crack tip B, as it does when the load is directed in the opposite direction of the crack. The values of the crack extension angle θ_2 for the case of Figure 2.20 are shown in Figure 2.21.

(c) *The cracked plate subjected to a localized moment*: For the case of Figure 2.13 when a localized moment M acts at the point K, the values of the stress intensity factors k_{1B} and k_{2B} are expressed by the following equations [16–18]:

$$\begin{aligned} k_{1B}(a;x,y) &= [\mathrm{Re}\,\phi(z)]k_0 \\ k_{2B}(a;x,y) &= -[\mathrm{Im}\,\phi(z)]k_0 \end{aligned} \tag{2.28a}$$

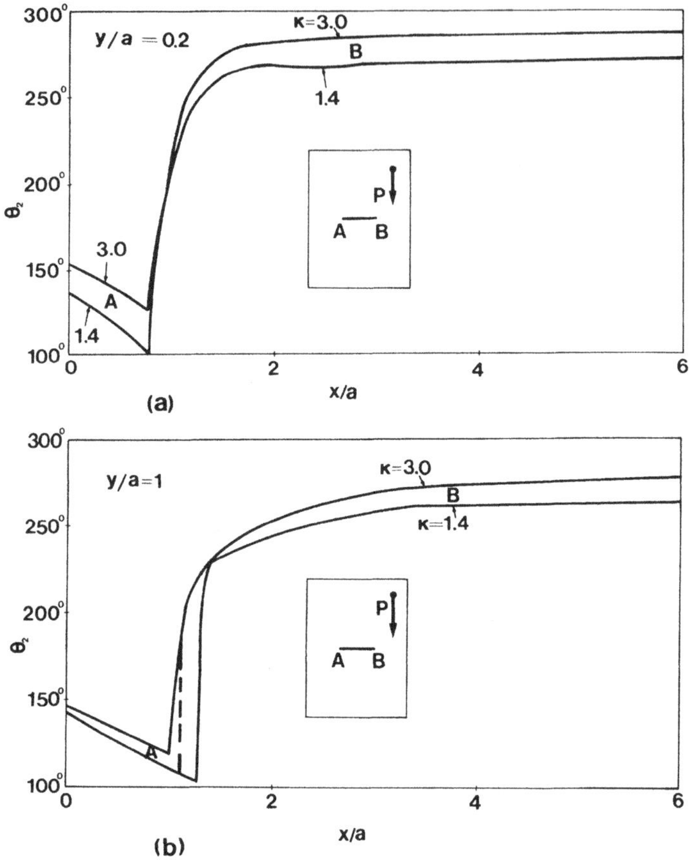

Figure 2.21. Values of the crack extension angle θ_2 for the case of Figure 2.20.

with:

$$\phi(z) = \frac{ia^2}{(\bar{z}-a)(\bar{z}^2-a^2)^{1/2}} \tag{2.28b}$$

and

$$k_0 = \frac{M}{2\pi a^{3/2}} \,. \tag{2.28c}$$

In these equations i is the imaginary unit, $\bar{z} = x - iy$ and Re and Im represent the real and imaginary parts of the corresponding function, respectively.

Due to the symmetry relations the values of the stress intensity factors k_{1A} and k_{2A} for tip A are given by:

$$\begin{aligned} k_{1A}(-a;x,y) &= -k_{1B}(a;-x,y) \\ k_{2A}(-a;x,y) &= k_{2B}(a;-x,y). \end{aligned} \tag{2.29}$$

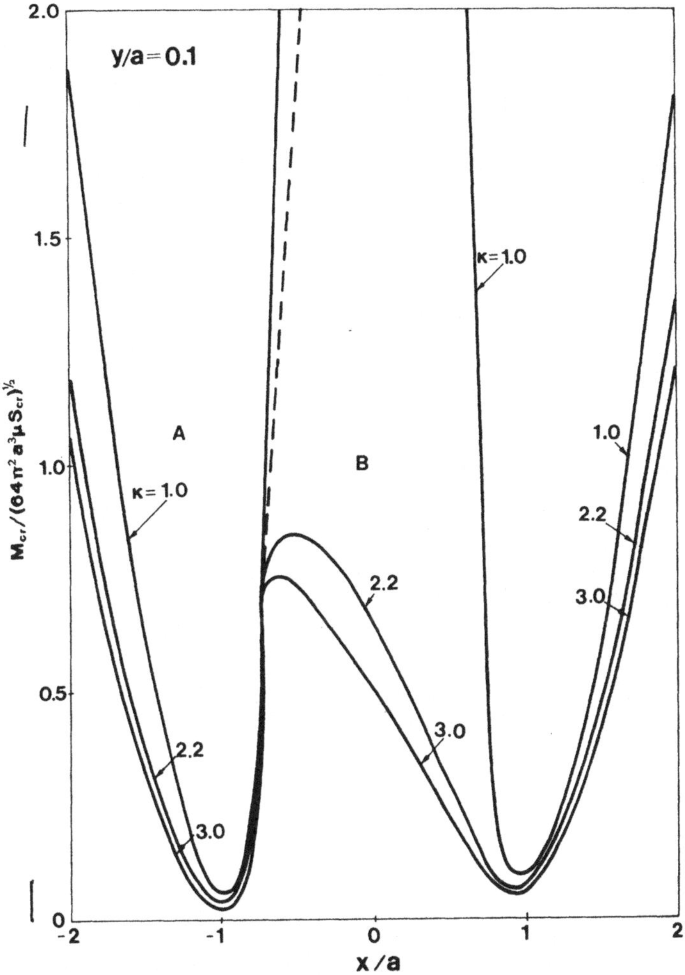

Figure 2.22. Variation of the normalized critical moment for crack extension $M_{cr}/(64\pi^2 a^3 \mu S_{cr})^{1/2}$ versus x/a for $y/a = 0.1$. Regions where crack extension starts from either of the tips A or B are separated by dotted lines.

Proceeding as in the previous cases of a concentrated force, the critical value of the applied moment M_{cr} for crack extension is determined. Figures 2.22 and 2.23 present the variation of the quantity $M_{cr}/(64\pi^2 a^3 \mu S_{cr})^{1/2}$ versus the distance x/a along the crack axis $(-2 \leqslant x/a \leqslant 2)$ for the values of the perpendicular distance y/a of the point of application of the moment M from the crack axis equal to 0.1 and 1.0. The material constant κ is assigned the values 1.0, 2.2 and 3.0. In both these figures the region where crack extension starts from tip A is separated by dotted lines from the region where it starts from tip B. The values of the crack extension angle θ_0 for the cases of Figures 2.22 and 2.23 are displayed in Figures 2.24 and 2.25.

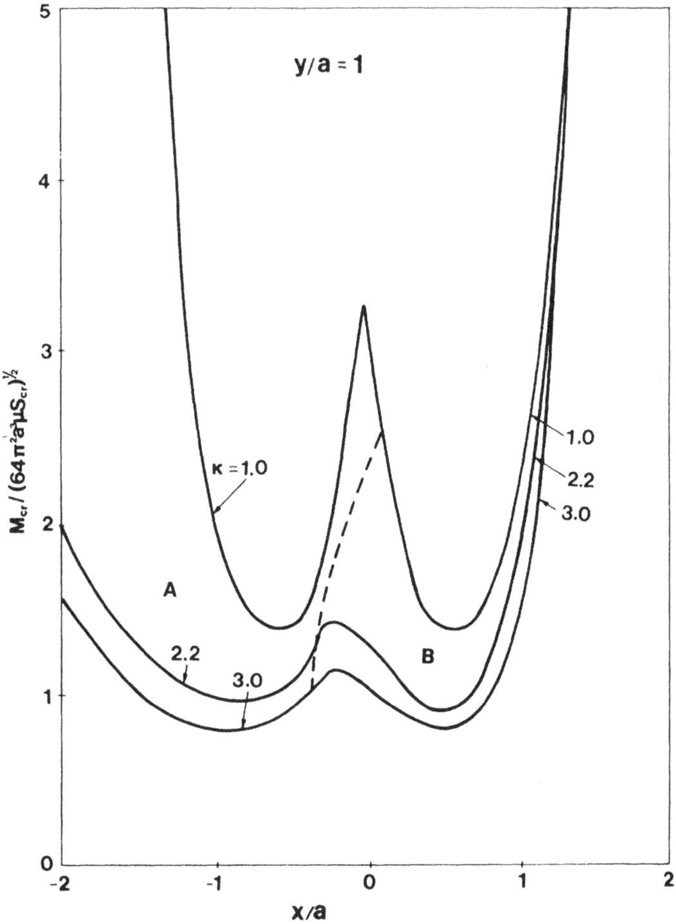

Figure 2.23. As in Figure 2.22 for $y/a = 1.0$.

From the above analysis of a cracked plate subjected to a concentrated moment at an arbitrary point the following conclusions concerning the variation of the critical moment required for crack extension from the more vulnerable crack tip, as well as the corresponding crack extension angle can be derived:

(1) For movement of the applied moment on lines parallel to the crack axis and from negative x coordinates, fracture of the cracked plate starts from the negative crack tip and continues up to a limiting value of (x/a) at which fracture starts from the positive crack tip. Table 2.3 shows the limiting values of x/a at which transition of crack extension from the negative to the positive crack tip takes place for various values of the distance y/a and the material constant κ.

(2) The critical moment for crack extension from the negative tip decreases abruptly as this tip is approached from negative x values, and takes its minimum value very close to this tip. More specifically, for small values of the distance y/a the minimum critical moments occur exactly above the negative tip, while for larger

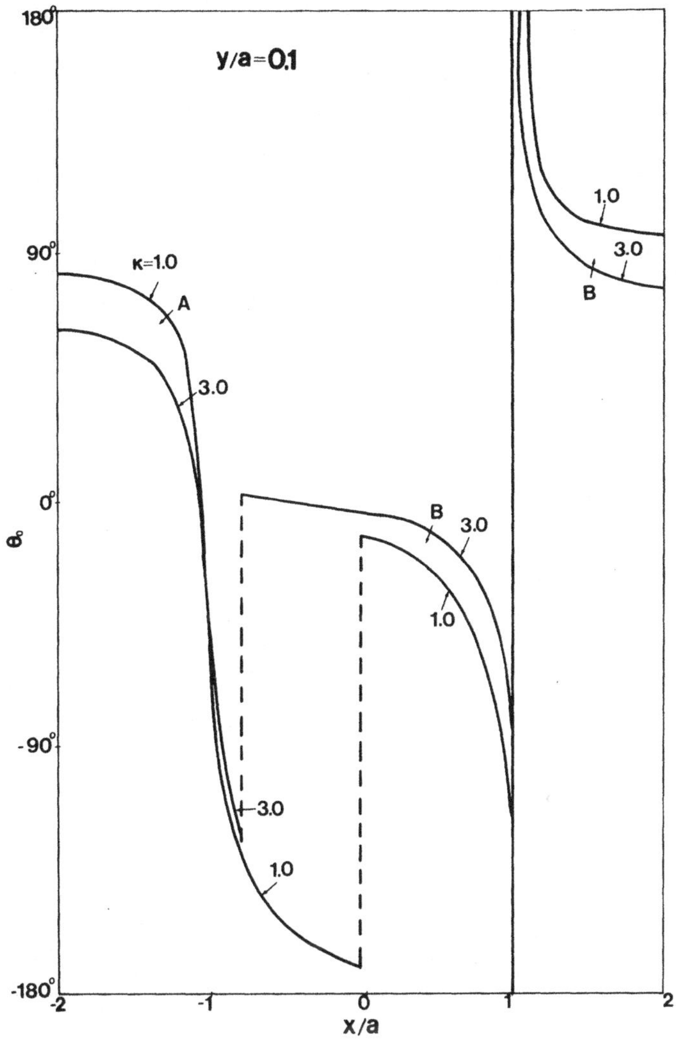

Figure 2.24. Values of the crack extension angle θ_0 from either of the tips A or B for the case of Figure 2.22.

Table 2.3. Critical values of the distance (x/a) for transition of crack extension from the negative to the positive tip.

	(y/a)				
κ	0.1	0.2	0.4	1.0	2.0
1.0	0	0	0	0	0
1.4	−0.67	−0.37	−0.30	−0.15	−0.10
2.2	−0.69	−0.65	−0.38	−0.33	−0.30
3.0	−0.78	−0.67	−0.42	−0.39	−0.35

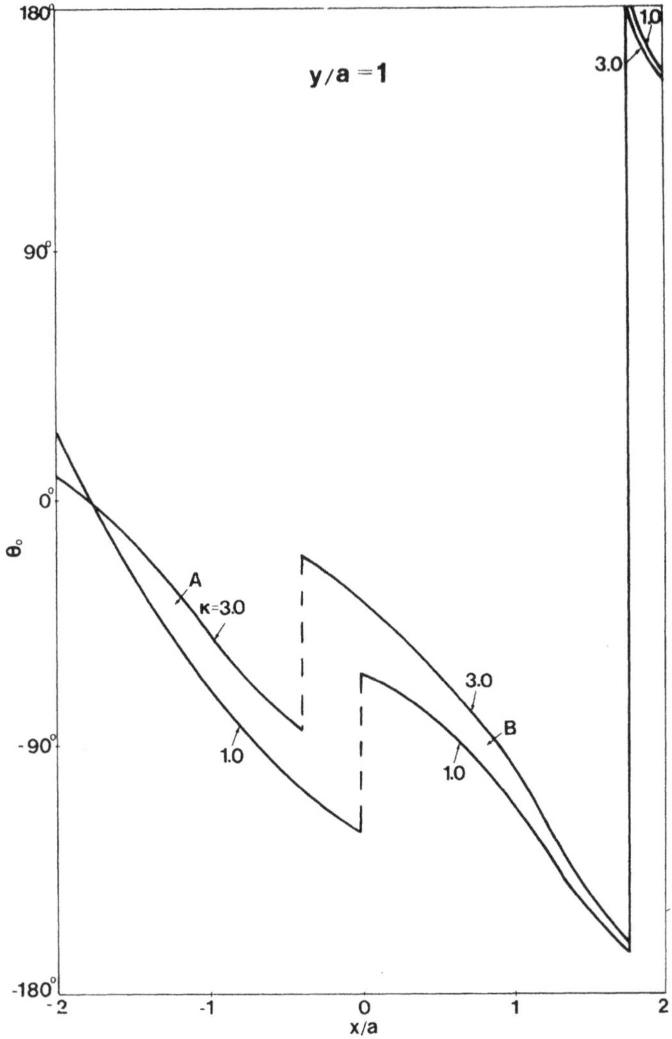

Figure 2.25. As in Figure 2.24 for $y/a = 1.0$.

values of y/a these positions are displaced toward the crack center. A displacement toward the crack center always takes place as κ decreases. After this point the critical moments increase and reach a local maximum in the region between the two crack tips. The same trend occurs with respect to the positive crack tip, with a similar tendency of the local minimum to be displaced toward the crack center. Comparing the two minima of M the minimum that occurs close to the negative tip is smaller than the other one. The critical value M_{cr} of the applied moment clearly depends on the value of the Poisson's ratio ν of the plate, decreasing as ν decreases.

(3) The branch of the crack formed after its extension from either of the tips moves in a clockwise direction as the point of application of the moment moves from negative to positive values.

For more details on the developments of this section the interested reader is referred to references [19, 20].

References

[1] Sih, G.C. and Liebowitz, H., Mathematical theories of brittle fracture, *Fracture*, Vol. II, edited by H. Liebowitz, Academic Press, New York, pp. 67–190 (1968).

[2] Sih, G.C., Paris, P.C. and Erdogan, F., Crack tip stress-intensity factors for plane extension and plate bending problems, *Journal of Applied Mechanics*, 29, pp. 306–312 (1962).

[3] Erdogan, F. and Sih, G.C., On the crack extension in plates under plane loading and transverse shear, *ASME Journal of Basic Engineering*, 85, pp. 519–527 (1963).

[4] Pook, L.P., The effect of crack angle on fracture toughness, *Journal of Engineering Fracture Mechanics*, 3, pp. 205–218 (1966).

[5] Hoek, E. and Bieniawski, Z.T., Fracture propagation mechanics in hard rock, *Technical Report – Rock Mechanics Division*, South African Council for Scientific and Industrial Research (1965).

[6] Sih, G.C., Some basic problems in fracture mechanics and new concepts, *Engineering Fracture Mechanics*, 5, pp. 365–377 (1973).

[7] Sih, G.C., Energy-density concept in fracture mechanics, *Engineering Fracture Mechanics*, 5, pp. 1037–1040 (1973).

[8] Sih, G.C., A special theory of crack propagation, *Methods of Analysis and Solutions of Crack Problems, Mechanics of Fracture 1*, edited by G.C. Sih, Noordhoff International Publishing, Leyden, pp. XXI–XLV (1973).

[9] Sih, G.C. and Kipp, M.E., Discussion on 'Fracture under complex stress', by J.G. Williams and P.D. Ewing, *International Journal of Fracture*, 10, pp. 261–265 (1974).

[10] Sih, G.C., Strain-energy-density factor applied to mixed mode crack problems, *International Journal of Fracture*, 10, pp. 305–321 (1974).

[11] Wilson, W.K., Numerical method for determining stress intensity factors of an interior crack in a finite plate, *Journal of Basic Engineering*, 93, pp. 685–690 (1971).

[12] Bowie, O.L., Solution of plane crack problems by mapping technique, *Methods of Analysis and Solutions of Crack Problems, Mechanics of Fracture 1*, Edited by G.C. Sih, Nordhoff International Publishing, Leyden, pp. 1–55 (1973).

[13] Wilson, W.K., On combined mode fracture mechanics, *Westinghouse Research Laboratories*, Research Report 69-1E7-FMECH-R1, Pittsburgh (1969).

[14] Gdoutos, E.E., Finite width effects on the crack extension angle, *International Journal of Fracture*, 15, pp. R111–R114 (1979).

[15] Gdoutos, E.E., The influence of specimen's geometry on the crack extension angle, *Engineering Fracture Mechanics*, 13, pp. 79–84 (1980).

[16] Erdogan, F., On the stress distribution in plates with collinear cuts under arbitrary loads, *Proceedings of the Fourth U.S. National Congress of Applied Mechanics*, pp. 547–553 (1962).

[17] Sih, G.C., Application of Muskhelishvili's method to fracture mechanics, *Transactions of the Chinese Association for the Advancement of Sciences*, pp. 2–10 (1962).

[18] Rooke, D.P. and Cartwright, D.J., Compedium of Stress Intensity Factors, London, Hillingdon Press, Uxbridge, pp. 36–61 (1976).

[19] Gdoutos, E.E., Fracture phenomena in a cracked plate subjected to a concentrated load, *Engineering Fracture Mechanics*, 14, pp. 323–335 (1981).

[20] Gdoutos, E.E., Fracture of cracked plates under localised moments, *International Journal of Mechanical Sciences*, 23, pp. 121–128 (1981).

3

Branched cracks

3.1. Introduction

Branched cracks are frequently encountered in fracture mechanics problems. A critical constant velocity together with a critical stress intensity factor constitute the necessary criteria for crack branching. Branching of propagating cracks has been observed during the rapid unstable fracture of various brittle materials as well as in stress corrosion cracking of high-strength steels. Branched cracks have also been observed when a single crack is subjected to a combined mode-one and mode-two loading, as seen in the preceding chapter. Furthermore, cracks with complicated geometry are usually found in microscopic cracking studies.

In all these cases interaction effects between the various crack branches take place. It is, therefore, essential to determine the particular crack tip from which rapid extension of the branched crack takes place under given applied loads. A number of studies aimed at determining the stress intensity factors at the tips of branched cracks have appeared in the literature [1–10]. These studies concern symmetric or asymmetric branched cracks, bent cracks and crack branches emanating from cracks with finite root radii.

It is the objective of this chapter to study the growth characteristics of various shapes of branched cracks existing in an elastic plate which is subjected to some simple types of loading. Cases of symmetrically and asymmetrically branched cracks, the bent crack and crack branches emanating from an elliptical crack will be considered. The strain energy density criterion is used to determine the more vulnerable branch of the crack from which brittle fracture of the plate takes place as well as the critical values of the applied loads. Furthermore, the initial fracture path from the tip of the critical branch is obtained. From this study the severity of a branched crack existing in a plate as compared to a single crack of the same overall length can be easily established. Furthermore, comparisons regarding the magnitude of the critical load for extension of various shapes of branched cracks are made.

3.2. The symmetrically branched crack

This type of crack branching is often encountered in fast fracture and stress corrosion cracking. The cracks may be either doubly or simply symmetrically branched (Figure 3.1). The cracks are embedded in an infinite elastic isotropic plate subjected to a uniform uniaxial tensile stress, either perpendicular or parallel to the main crack for the case of the doubly symmetric branched crack, or perpendicular to the main crack for the simply symmetric branched crack. These three cases will be studied individually below.

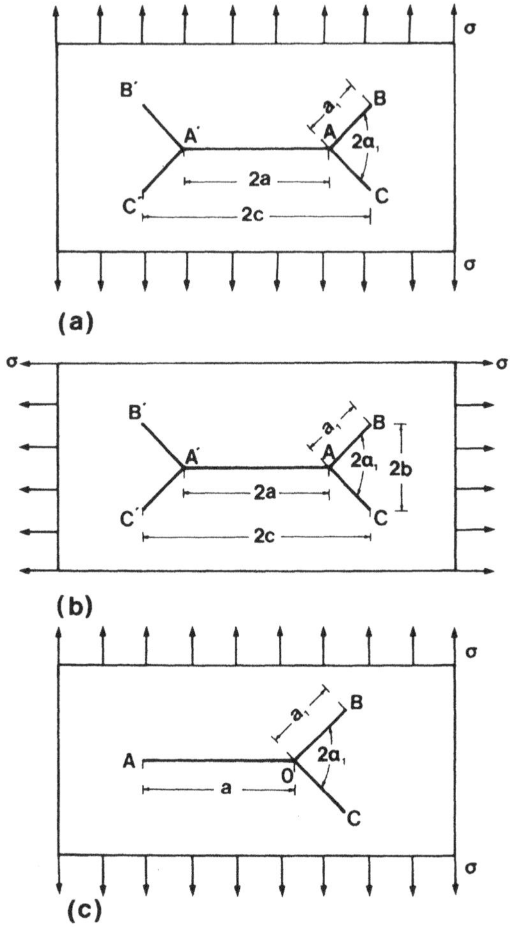

Figure 3.1. Geometry of a doubly symmetric branched crack embedded in an infinite isotropic elastic plate subjected to a tensile stress perpendicular (a) or parallel (b) to the main branch. The simply symmetric branched crack loaded perpendicularly to the main branch (c).

a. *The doubly symmetric branched crack with the applied load perpendicular to the main crack* (Figure 3.1a): Consider the crack of Figure 3.1a composed of the main branch $A'A$ of length $2a$ and the four equal branches, $AB, AC, A'B', A'C'$ of length a_1 symmetrically located to the crack $A'A$ at an angle α_1. The crack is embedded in an infinite elastic isotropic body which is subjected to a uniform uniaxial tensile stress σ at infinity perpendicular to the main crack. It is assumed that the dimensions of the body in the z-direction are very large, so that plane strain conditions prevail in the area of the crack. This problem has been considered by Vitek [9] who found the values of the opening-mode k_1 and sliding-mode k_2 stress intensity factors at the tips of the branches. Due to the presence of symmetry, k_1 and k_2 are the same for all four branches.

The general equations of the strain energy density criterion for the two-dimensional

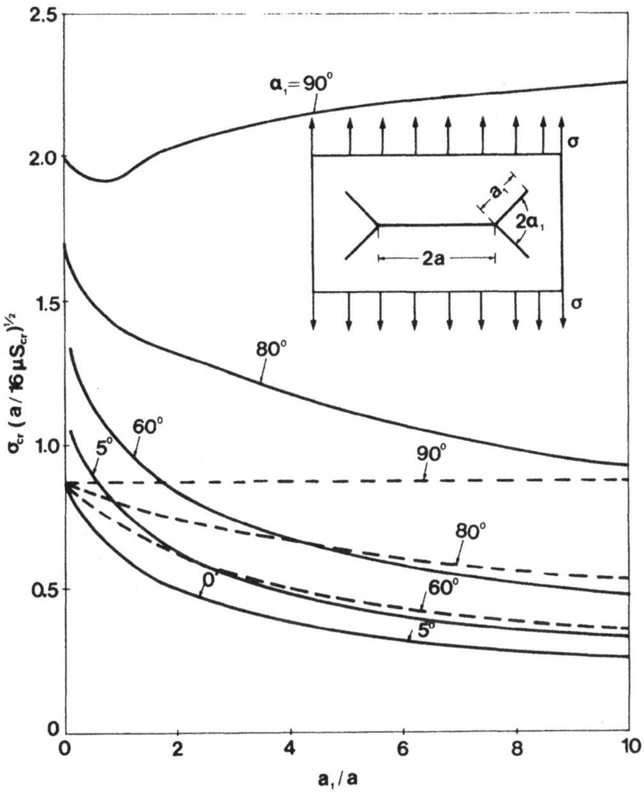

Figure 3.2. Variation of the quantity $\sigma_{cr}(a/16\mu S_{cr})^{1/2}$ versus the normalized branch length a_1/a for the doubly symmetric branched crack of Figure 3.1a for $\alpha_1 = 0°, 5°, 60°, 80°$ and $90°$. Dotted curves indicate the values of the above quantity for the equivalent crack of length $2(a + a_1 \cos\alpha_1)$.

crack problem developed in Section 2.2 (page 11) were used to determine the brittle fracture characteristic quantities of the plate. Thus the values of k_1 and k_2 for each particular geometrical configuration of the crack were introduced into equation (2.6a). Solution of the resulting equation for a given value of the material constant κ, under restrictions imposed by inequality (2.6b), gives the value of the angle θ_0 which each of the four propagating branches forms with its initial direction. When the values of k_1, k_2, κ and θ_0 are introduced into equations (2.4) and (2.3) the minimum value S_{min} of the strain energy density factor S is obtained. Equating S_{min} to the critical strain energy density factor S_{cr}, which is a material constant, the critical value σ_{cr} of the applied stress σ for crack extension is obtained. Due to the significance of this type of crack branching in practice the variation of the stress σ_{cr} with the length and the inclination of the branches will be considered separately.

Figure 3.2 presents the variation of the non-dimensional quantity $\sigma_{cr}(a/16\mu S_{cr})^{1/2}$ versus the normalized branch length a_1/a for the values of the angle α_1 of the inclination of the branches with respect to the plane of the main crack equal to $0°$, $5°$, $60°$, $80°$ and $90°$. The material constant κ for all curves of figure is set equal to 1.667

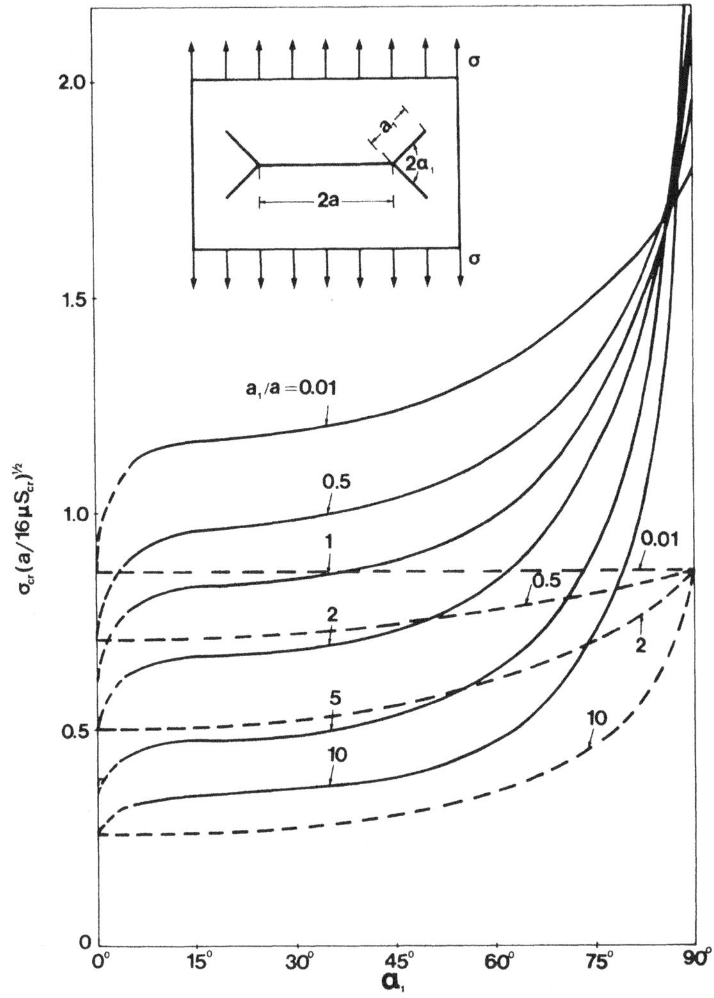

Figure 3.3. As in Figure 3.2 for the variation of the quantity $\sigma_{cr}(a/16\mu S_{cr})^{1/2}$ versus the branch inclination angle α_1 for $a_1/a = 0.01, 0.5, 1, 2, 5$ and 10.

(corresponding to a value of the Poisson's ratio ν of the material of the plate equal to 1/3 under plane strain conditions). The values of the same quantity $\sigma_{cr}(a/16\mu S_{cr})^{1/2}$ versus the angle $\alpha_1(0 \leqslant \alpha_1 \leqslant 90°)$ for $a_1/a = 0.01, 0.5, 1, 2, 5$ and 10 are presented in Figure 3.3.

From Figure 3.2 we observe that for a given angle α_1 of inclination of the branches the critical stress σ_{cr} decreases with the branch length a_1 for all angles α_1 except for $\alpha_1 = 90°$. In the latter case, when $a_1 > 0.52a$, σ_{cr} increases as a_1 increases. This is also evident from Figure 3.3, where it is shown that for angles α_1 close to 90° ($85° < \alpha_1 < 90°$) the stress σ_{cr} increases with the length of the branches a_1 after a_1 reaches a critical length. It is therefore concluded that for the H-shaped crack loaded perpendicularly to the main crack, and for branch lengths greater than a critical length, the stress σ_{cr} for extension of the branches increases with the branch length. For a given value of the length of the branches a_1 the critical stress σ_{cr} increases as the inclination angle of the branches α_1 increases (Figure 3.3). Thus the more severe doubly symmetric branched cracks are those with long branches under small angles of inclination with respect to the plane of the main crack.

In order to obtain an estimate of the effect that a branched crack has on the strength of a body as compared to a line crack we define the equivalent crack of length $2c$ equal to $2(a+a_1\cos\alpha_1)$ (Figure 3.1a). The values of the critical stress of brittle fracture of a body weakened by the equivalent crack are presented in Figures 3.2 and 3.3 by dotted lines. It is observed that the critical stress for the equivalent crack is always smaller than for the branched crack. As is shown in Figure 3.2 the difference between the two stresses increases as the angle of inclination of the branches increases (thus for $\alpha_1 = 5°$ the critical stress for the branched crack is greater by a factor 1.3 of this stress for the equivalent crack, while for $\alpha_1 = 90°$ this factor becomes 2.5). It is thus concluded that this type of crack branching always retards propagation as compared to the line crack. The critical stress of the branched crack was also compared with the corresponding stress for the main crack of length $2a$ without branches. The value of the latter is presented in Figure 3.2 by the dotted curve corresponding to $\alpha_1 = 90°$ and in Figure 3.3 approximately by the dotted curve for $a_1/a = 0.01$ (the value of the quantity $\sigma_{cr}(a/16\mu S_{cr})^{1/2}$ for this case is equal to 0.866). From Figure 3.2 it is observed that the critical stress for the branched crack is always greater than for the main crack for $\alpha_1 = 80°$ and $90°$, as well as for small branch lengths for $\alpha_1 = 5°$ and $60°$. Thus for $\alpha_1 = 5°$ and $a_1 < 0.60a$ and for $\alpha_1 = 60°$ and $a_1 < 0.83a$ the critical stress of the branched crack is greater than for the main crack. The same result is shown in Figure 3.3 for crack branches of given lengths whose angle α_1 of inclination varies. Thus, for $a_1 = 0.01a$ the critical stress for the branched crack is always greater than for the main crack. This also holds true for $a_1 = 0.5a$ and $\alpha_1 > 3.5°$, for $a_1 = a$ and $\alpha_1 > 37.5°$, for $a_1 = 2a$ and $\alpha_1 > 62°$, for $a_1 = 5a$ and $\alpha_1 > 68°$ and for $a_1 = 10a$ and $\alpha_1 > 79°$. It is thus concluded that for branches of either small lengths or large inclination angles the existence of the branches slows the propagation.

The values of the angle that the fracture path starting from the tip B of the branch AB forms with the line AB are shown in Figure 3.4. Angle θ_0 is the polar angle in a Cartesian coordinate system with its origin at the tip B of the branch AB and the positive x-axis having the direction of AB and oriented outside of it. Thus, positive angles θ_0 mean that the branch AB extends in a counterclockwise direction, while negative angles θ_0 stand for a clockwise propagation of AB. In Figure 3.4 it is seen that for small values of the branch angle α_1 and depending on the length a_1 of the branches, the angle θ_0 is positive. Thus, θ_0 is positive for $\alpha_1 < 22°$ and $a_1 = 0.01a$, for $\alpha_1 < 12.5°$ and $a_1 = 0.5a$ and for $\alpha_1 < 9°$ and $a_1 = 10a$. For these cases the branches tend to propagate with increasing angle, which means that they have the tendency to repel each other. For the remaining values of α_1 the branches propagate with decreasing angle; that is, they tend to attract each other. For those particular values of the angle α_1 and the length a_1 defined above, the crack propagation angle θ_0 is zero and therefore the branches maintain their original direction. Thus, for branches with small angle α_1 and long lengths a_1 or larger angles α_1 and shorter lengths a_1 the branch angle increases during crack extension. This result agrees with that obtained by Kalthoff

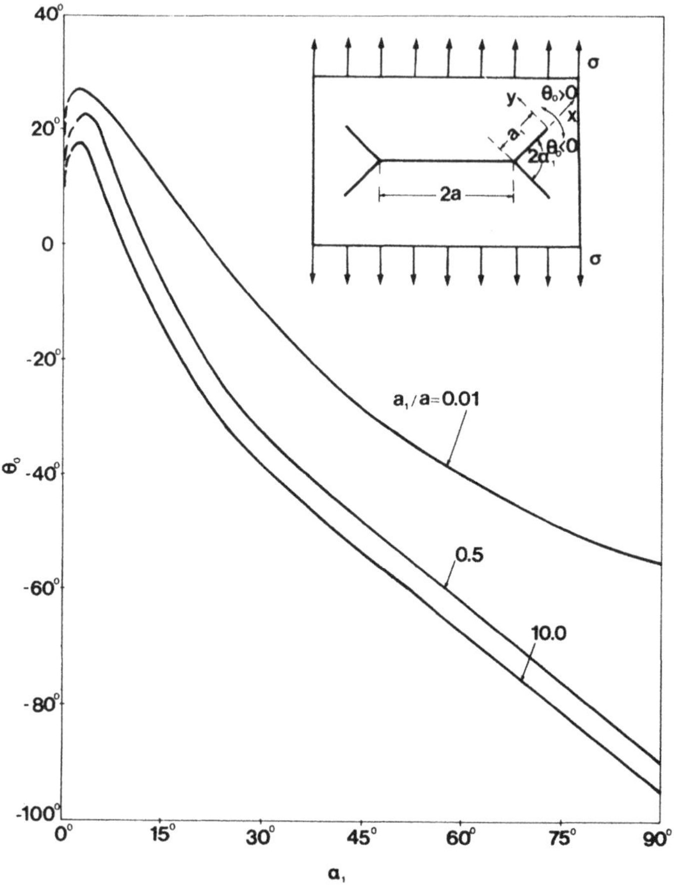

Figure 3.4. Values of the angle θ_0 that the fracture path starting from the tip B of the branch AB forms with the line AB for the crack of Figure 3.1a.

[11], who found that for the case of two symmetric branches emanating from the boundary of a plexiglas plate the critical branch angle at which the branches retain their original direction is approximately 14°. Calculating the angle formed by the fracture path originating from the tip of the branches and the direction of the applied stress we find that this angle lies in the interval (55°, 73°) for $a_1 = 0.01a$, in the interval (63°, 93°) for $a_1 = 0.5a$ and in the interval (66°, 98°) for $a_1 = 10a$. Thus, the branches extend in a direction which approaches that perpendicular to the applied tensile stress as the branch length increases.

b. *The doubly symmetric branched crack with the applied load parallel to the main crack (Figure 3.1b)*: This problem has been studied by Kitagawa and Yuuki [10], who gave the values of the k_1 and k_2 stress intensity factors at the tips of the branches. Working as in the previous case, the critical values of the applied tensile stress for crack extension and the corresponding fracture angles are determined. Figure 3.5 presents the variation of the quantity $\sigma_{cr}(c/16\mu S_{cr})^{1/2}$ versus the normalized branch crack

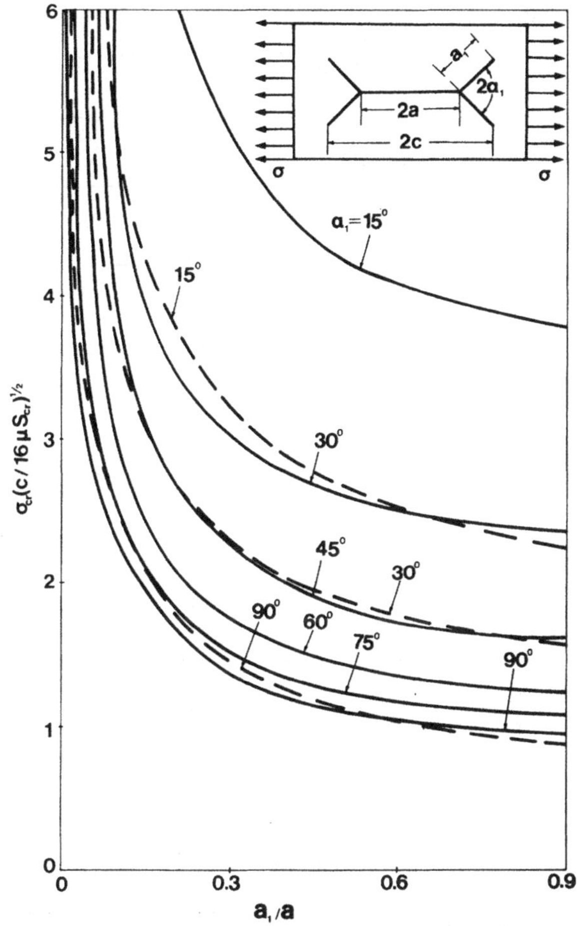

Figure 3.5. Variation of the quantity $\sigma_{cr}(c/16\mu S_{cr})^{1/2}$ versus the normalized branch length a_1/a for the doubly symmetric branched crack of Figure 3.1b for $\alpha_1 = 15°, 30°, 45°, 60°, 75°$ and $90°$. Dotted curves indicate the values of the above quantity for the equivalent crack of length $2(a + a_1 \cos\alpha_1)$.

length to the half length of the main crack a_1/a for the values of the branch angle α_1 equal to 15°, 30°, 45°, 60°, 75° and 90°. The material constant κ was assigned the value 1.8. We observe from this figure that the critical stress σ_{cr} decreases as either the branch length a_1 or the branch angle α increases. For $\alpha_1 = 0°$ the branched crack degenerates to a line crack parallel to the applied stress and therefore the critical fracture stress σ_{cr} tends to infinity. Furthermore, for $a_1 = 0$ the branched crack becomes equal to a straight crack of length $2a$ parallel to the applied stress, and the critical stress tends to infinity.

In order to obtain a comparison of the influence of this type of crack branching on the strength of a body with a straight crack we define an equivalent crack by the projection of the branched crack in the direction perpendicular to the applied stress

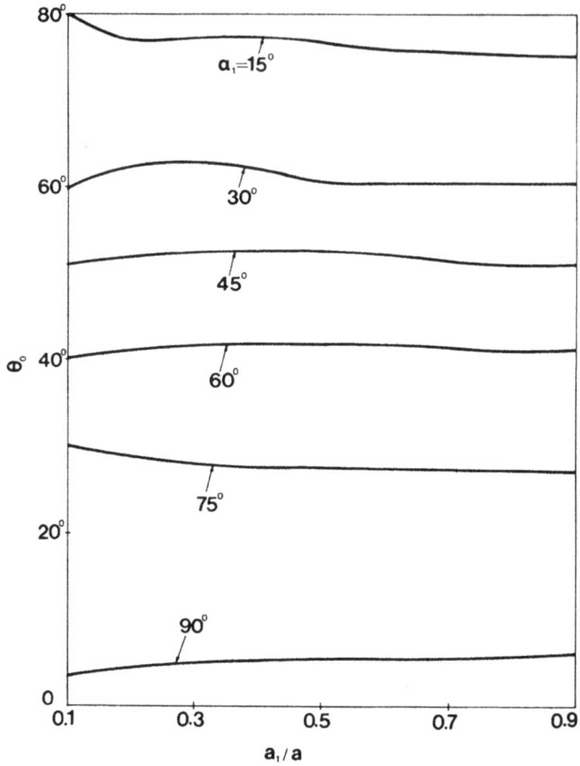

Figure 3.6. Values of the angle θ_0 that the fracture path originating from the tip B of the branch forms with the line AB for the crack of Figure 3.1b.

(length of the equivalent crack equal to $2a_1 \sin \alpha_1$). The values of the critical stress to propagate the equivalent crack are shown in Figure 3.5 by dotted lines. It is observed that the equivalent crack is always more damaging to the strength of the body than the branched crack, except for the H-shaped crack ($\alpha = 90°$) for $a_1 > 0.6a$. Thus, the strength of a body with this type of branched crack is always greater than that of a body with a straight crack equal to the projection of that crack on the direction normal to the applied stress. The strength of a body decreases as either the length of the branches or their inclination increases.

The values of the angle θ_0 formed by the fracture path originating from the tip B of the branch AB with the line AB are shown in Figure 3.6. It is observed that the angle θ_0 is positive and so the branches tend to propagate with increasing angle. In other words the branches repel each other during crack extension. We also observe that the angle θ_0 is almost independent of branch length. Calculating the angle ϕ that the fracture path forms with the direction of the applied stress we find that for all branch lengths and inclinations ϕ varies in the interval $(90°, 104°)$. That is, the fracture path starting from the tips of the branches is approximately perpendicular to the direction of the applied tensile stress.

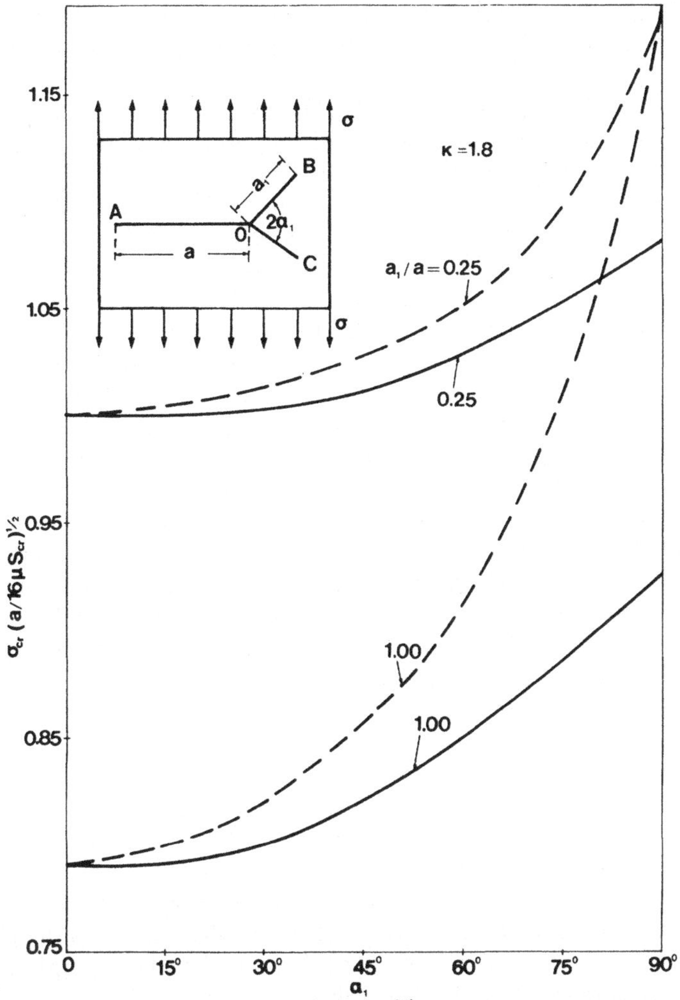

Figure 3.7. Variation of the quantity $\sigma_{cr}(a/16\mu S_{cr})^{1/2}$ versus the branch inclination angle α_1 for the simply symmetric branched crack of Figure 3.1c for $a_1/a = 0.25$ and 1.0. Dotted curves indicate the values of this quantity for the equivalent crack of length $(a + a_1 \cos\alpha_1)$.

c. *The simply symmetric branched crack (Figure 3.1c)*: Consider the simply symmetric branched crack of Figure 3.1c composed of the main branch *OA* of length a and the branches *OB* and *OC* of length a_1 inclined under angles α and $-\alpha$ to the main branch. The crack is embedded in an infinite isotropic elastic plate which is loaded by a uniform uniaxial stress σ perpendicular to the direction of the main branch *OA*. For this case with a progressively increasing applied stress σ we will determine the critical value σ_{cr} of σ for which brittle fracture of the plate takes place by propagating the more vulnerable crack branch.

This problem has been considered by Theocaris and Ioakimidis [5] who found the values of the k_1 and k_2 stress intensity factors at the tips of the branches. The sliding mode stress intensity factor k_2 for the tip A is equal to zero due to the symmetry of the crack shape and loading with respect to the main branch OA, which is aligned perpendicularly to the direction of the applied stress. Using the equations of the strain energy density criterion for the general two-dimensional crack problem for the tip B the critical value σ_{cr}^B of the applied stress for possible brittle fracture of the plate starting from B was determined. Also, the critical stress σ_{cr}^A for possible brittle fracture originating from the tip A was calculated. It is evident that the actual fracture stress of the plate is the minimum of the two σ_{cr}^A and σ_{cr}^B and that the fracture would take place by propagating the tip with the minimum required critical stress.

Following this procedure it was found that brittle fracture of the plate always takes place by propagating the main branch OA. Due to the symmetry of the problem, the branch OA extends in its own plane. Figure 3.7 presents the variation of the quantity $\sigma_{cr}(a/16\mu S_{cr})^{1/2}$ versus the branch angle α_1 for branch lengths a_1 equal to a or $0.25a$. Both curves shown in the figure correspond to $\kappa = 1.8$. It is observed that the critical stress σ_{cr} increases as the angle α_1 of branch inclination increases. We also see that σ_{cr} decreases as the length of the branches increases. The critical stress σ_{cr} was compared with the critical stress of the plate weakened by the equivalent crack, which is a straight crack of length $(a + a_1 \cos\alpha_1)$ defined by the projection of the branches in the direction of the main crack. The values of the critical stress for the equivalent crack were plotted in Figure 3.7 by dotted lines. It is observed that, unlike the previous cases of the doubly symmetric branched crack, the critical stress for the branched crack is always smaller than for the equivalent crack.

3.3. The asymmetrically branched crack

Three types of asymmetrically branched cracks will be considered. In the first, two equally inclined branches of different lengths emanate from both tips of the main branch (Figure 3.8a); in the second, two branches of unequal inclination and length emanate from one tip of the main branch (Figure 3.8b); and in the third an inclined branch emanates from a point of the main branch other than its tips (Figure 3.8c). These three cases will be studied individually.

a. *The asymmetrically branched crack with four equally inclined branches of different lengths (Figure 3.8a)*: Consider the crack of Figure 3.8a composed of the main branch $A'A$ of length $2a$ and branches AB, $A'B'$ and AC, $A'C'$ of lengths a_1 and a_2, which are inclined at angles of $45°$ and $-45°$ to the main branch. The crack is embedded in an infinite isotropic elastic plate loaded by a uniform uniaxial stress σ perpendicular to the main crack. For this geometrical configuration with a Poisson's ratio ν of the plate equal to $1/3$, Vitek [9] gave the values of the k_1 and k_2 stress intensity factors for tips B and C of the branches AB and AC under conditions of plane strain. Using the equations of the strain energy density criterion with the given values of k_1 and k_2 for the tips B and C, the critical stresses σ_{cr}^B and σ_{cr}^C for possible crack extension from the tips B and C are determined. It is evident that the crack would extend from the tip which requires the smaller critical stress for propagation. Working in this way it is found that the crack always extends from tip B of the longer branch AB. The critical values σ_{cr} of the applied stress σ versus the ratio a_1/a_2 $(1 < a_1/a_2 < 10)$

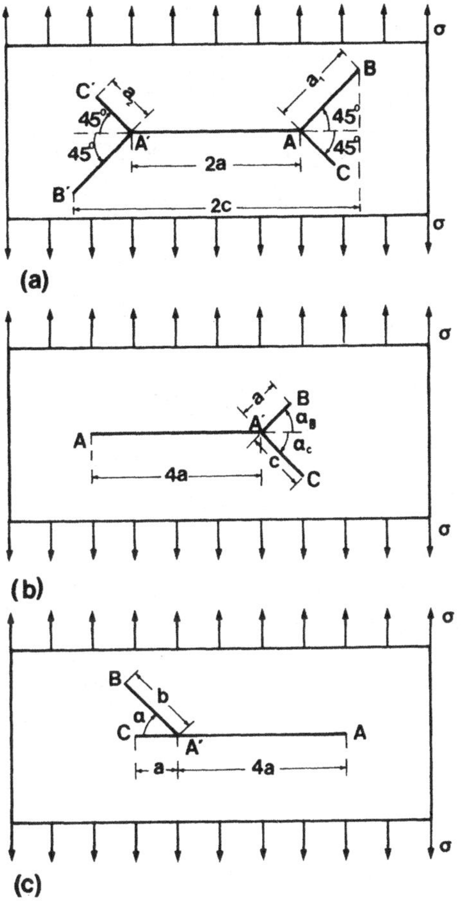

Figure 3.8. Geometrical configurations of asymmetric cracks.

for $a_2/a = 0.01, 0.1, 0.5$ and 10 are shown in Figure 3.9. It is observed that the critical stress σ_{cr} decreases as either of the branches of the crack increases. In order to compare this case with the case of a doubly symmetrical branched crack whose branches are equal to the longer branch of the asymmetric crack, the critical values σ_{cr} of the stress σ for the doubly symmetrical branched crack are shown in Figure 3.9 by dotted lines. It is observed that the stress for the asymmetric crack is always smaller than for the symmetric crack. This can be attributed to the fact that when the crack branches are equal the energy induced in the body is equally distributed between the branches, while when the branches are unequal most of this energy is absorbed by the longer branch and therefore smaller external energy is required for crack extension.

The values of the angle θ_0 formed by the fracture path originating from the tip B of the branch AB with the direction of AB are displayed in Figure 3.10. As in the case of the doubly symmetric branched crack the angle θ_0 is always negative. This means that the crack tends to propagate with decreasing branch angle. The values of the angle

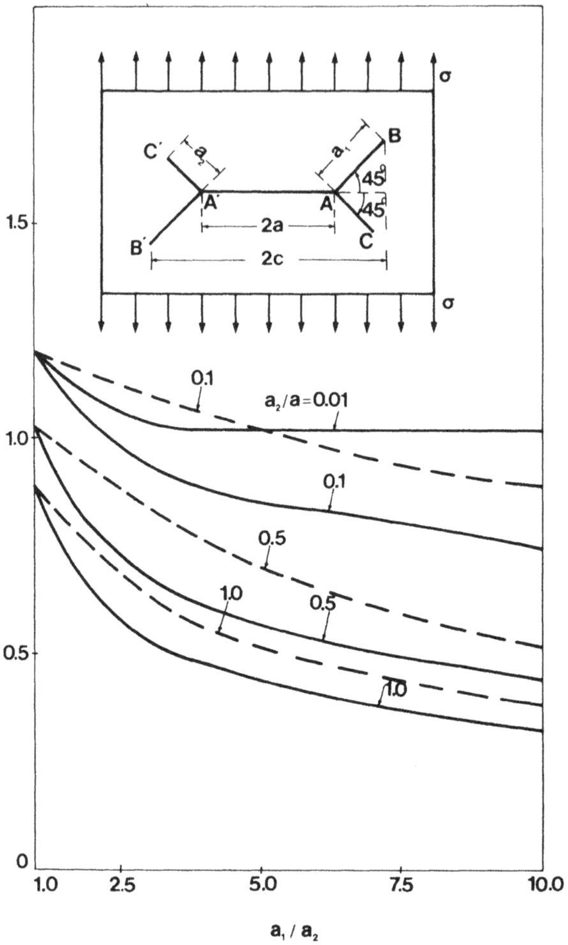

Figure 3.9. Variation of the quantity $\sigma_{cr}(a/16\mu S_{cr})^{1/2}$ versus a_1/a_2 for the crack of Figure 3.8a for $a_2/a = 0.01$, 0.1, 0.5 and 1.0. Dotted curves indicate the values of the above quantity for the doubly symmetric branched crack with branches equal to a_1.

formed by the propagating branch with the direction of the applied stress for all geometrical configurations of Figure 3.8a lie in the interval (73°, 98°). We see that the propagating branch is rather close to the direction perpendicular to the applied stress.

b. *The asymmetrically branched crack with two unequal branches of different lengths (Figure 3.8b)*: This crack is composed of the main branch AA' of length $4a$ and the branches $A'B$ and $A'C$ of lengths a and c subtending to the main branch under angles α_B and α_C respectively. The plate is loaded by a uniform stress perpendicular to the main branch. This crack configuration has been studied by Theocaris [6] who gave the values of k_1 and k_2 for the tips B and C normalized to the stress intensity factor k_1 for the tip A. The latter can be approximated by the stress intensity factor of a straight line crack of length equal to the projection of the whole crack to the direction perpendicular to the applied stress. Using the equations of the strain energy density

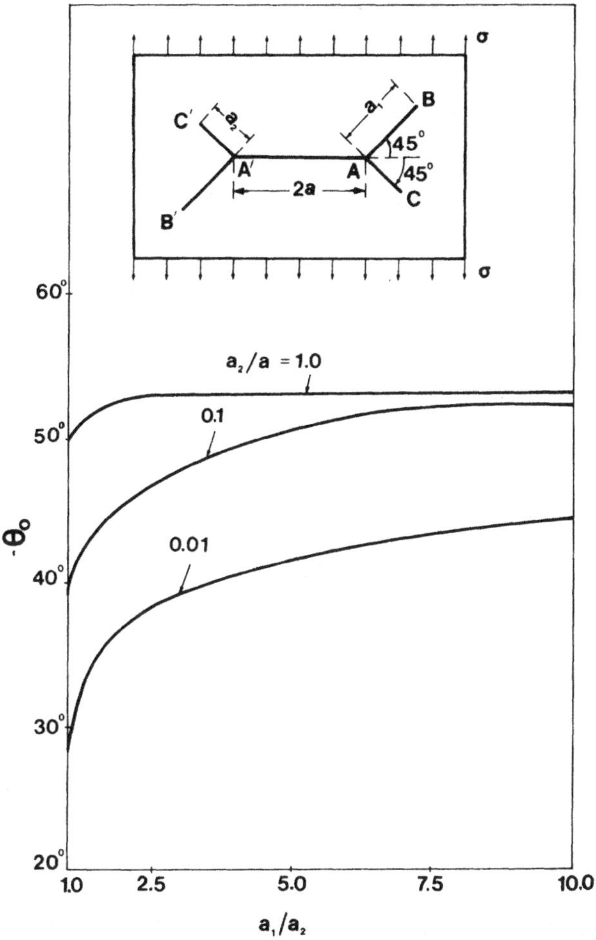

Figure 3.10. Values of the angle $-\theta_0$ that the fracture path originating from the tip B of the longer branch AB forms with the line AB for the crack of Figure 3.8a.

criterion with the above values of stress intensity factors, the critical stress for possible crack propagation from all three tips A, B and C of the crack are determined. The more vulnerable tip from which the crack extends corresponds to that with the smaller of the above three stresses. Using this procedure it was found that the crack always propagates from the tip A of the main crack AA'. This can be attributed to the fact that the external energy imposed in the crack is distributed between the tips of its two ends, in this case, between A and, B and C. When two branches ($A'B$, $A'C$) are formed from one crack end, the energy is shared between them and therefore the energy absorbed by the tip of the other crack end A is always larger than that absorbed by each of the two tips B and C. The values of the critical stress σ_{cr} for $\alpha_B = 30°$ and $\alpha_C = 15°$, $45°$, $60°$ and $90°$ are shown in Figure 3.11 by solid lines. The same curves for $\alpha_B = 60°$ are presented by dotted lines. It is observed that for a constant angle α_B the stress σ_{cr} decreases as the angle α_C decreases. Furthermore, this stress decreases as the length c of the branch $A'C$ increases.

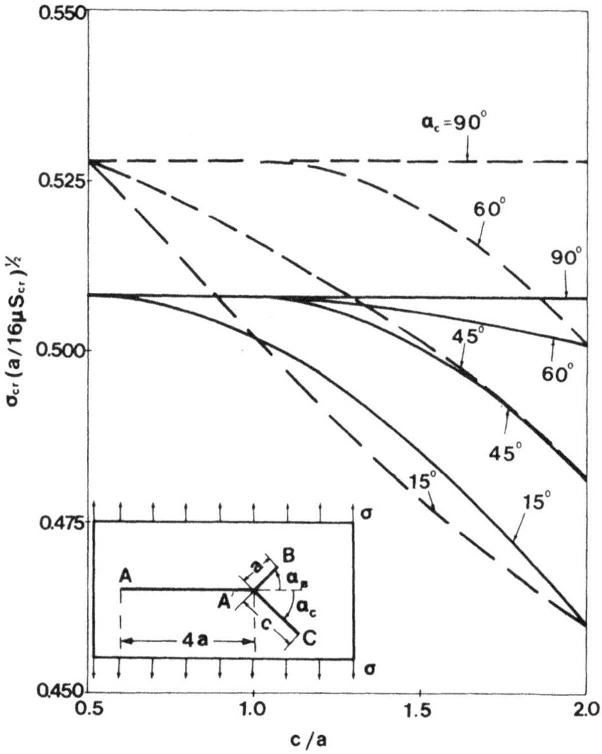

Figure 3.11. Variation of the quantity $\sigma_{cr}(a/16\mu S_{cr})^{1/2}$ versus c/a for the crack of Figure 3.8b for $\alpha_B = 30°$ and $\alpha_C = 15°, 45°, 60°$ and $90°$ (continuous curves) and for $\alpha_B = 60°$ and $\alpha_C = 15°$, $45°, 60°$ and $90°$.

c. *The asymmetrically branched crack with an inclined branch emanating from a point of the main crack other than its ends (Figure 3.8c).* For this crack configuration Theocaris [3] gave the values of the k_1 and k_2 stress intensity factors for tips B and C of branches $A'B$ and $A'C$. The value of k_1 for tip A can be approximated to that of a crack normal to the direction of the applied stress. Working as in the previous case it was found that the crack always propagates from tip A of the main branch AA'. The values of the critical stress σ_{cr} versus b/a for $\alpha = 15°$, $30°$ and $45°$ are shown in Figure 3.12. From this figure observations similar to those made in the previous case of the asymmetric crack can be made.

3.4. The bent crack

In this section the fracture behavior of a plate with a bent crack will be studied. Two types of bent cracks will be considered: the skew-symmetric bent crack composed of a main branch AA' and two inclined branches AB and $A'B'$ emanating from the ends A and A' of the main branch having equal lengths and subtending equal angles with AA'

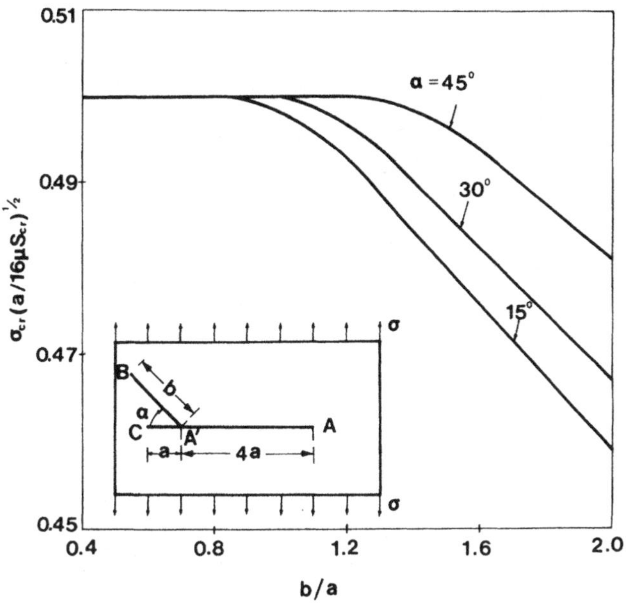

Figure 3.12. Variation of the quantity $\sigma_{cr}(a/16\mu S_{cr})^{1/2}$ versus b/a for the crack of Figure 3.8c for $\alpha = 15°, 30°$ and $45°$.

(Figure 3.13a, b); and the bent crack with two branches composed of a main branch and another inclined to the main and starting from one of its ends (Figure 3.13c). For the first case of the skew-symmetric bent crack two types of uniaxial uniform tensile load will be considered. One is normal to the main crack (Figure 3.13a) and the other subtends at an angle equal to 45° (Figure 3.13b). For the second case of the bent crack with two branches the plate is subjected to a uniaxial uniform tensile stress perpendicular to the main branch (Figure 3.13c).

a. *The skew-symmetric bent crack (Figure 3.13a, b)*: Consider the skew-symmetric bent crack $B'A'AB$ embedded in an infinite isotropic elastic plate subjected to a uniform uniaxial stress σ at infinity perpendicular to the main branch AA'. For this case under conditions of plane strain Vitek [9] calculated the values of the opening-mode k_1 and sliding-mode k_2 stress intensity factors for the tips B and B' of the inclined branches AB and $A'B'$. Introducing these values of k_1 and k_2 into the equations of the strain energy density criterion (Section 2.2, page 11) the critical value σ_{cr} of the applied stress σ for crack extension was determined. Figure 3.14 presents the variation of the quantity $\sigma_{cr}(a/16\mu S_{cr})^{1/2}$ versus the angle α_1 of inclination of the branches for $a_1/a = 0.01$, 0.1, 0.5 and 1.0 when the Poisson's ratio of the material of the plate is equal to 1/3. For $\alpha_1 = 0$ the value of the quantity $\sigma_{cr}(a/16\mu S_{cr})^{1/2}$ corresponds to that obtained for a line crack of length $2(a + a_1)$ loaded in tension in a direction perpendicular to the plane of the crack. From Figure 3.14 we observe that as in the cases of the symmetrically branched cracks, the critical stress σ_{cr} increases as the angle α_1 of inclination of the branches increases. The stress σ_{cr} decreases as crack length increases for angles α_1 smaller than 60°. For angles α_1 in the interval (60°, 90°) however, this rule is not valid.

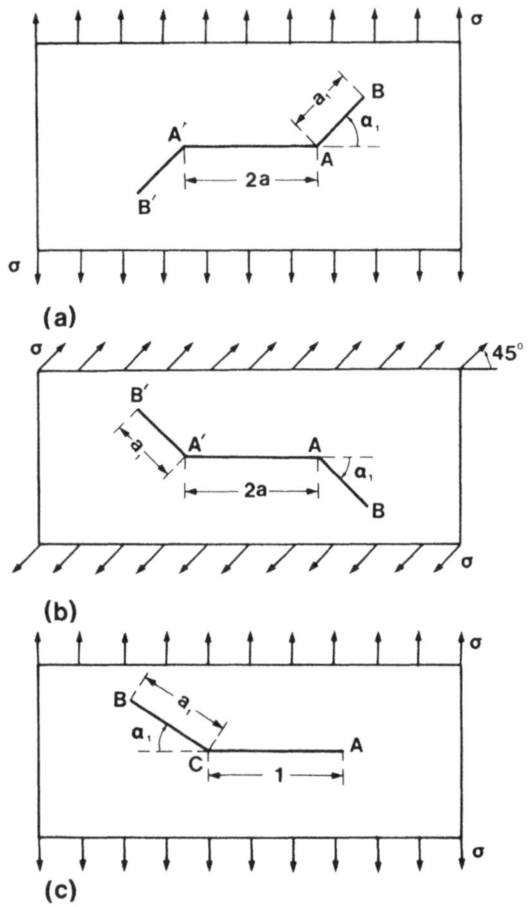

Figure 3.13. Geometrical configurations of bent cracks.

As in the case of the symmetrically branched crack, we define the equivalent crack as the line crack of length $2(a + a_1 \cos \alpha_1)$ defined by the projections of the inclined branches to the plane of the main branch. The values of the critical stress for the equivalent crack are given in Figure 3.14 by dotted lines. It is observed that, as in the doubly symmetric branched crack, the fracture stress of the branched crack is always greater than that of the equivalent crack.

The values of the angle $-\theta_0$ formed by the path of the propagating branch AB starting from the tip B with its initial direction AB are displayed in Figure 3.15. It is observed that the angle θ_0 is always negative; therefore, the branches propagate with decreasing angle and tend to become perpendicular to the direction of the applied stress. By calculating the angle ϕ that the propagating path of the branch AB forms with the direction of loading we find that this angle varies in the interval (56.7°, 90°) for $a_1 = 0.01a$, in (72.4°, 95.9°) for $a_1 = 0.1a$, in (88.9°, 99.4°) for $a_1 = 0.5a$ and in (81.6°, 100°) for $a_1 = a$. It is observed that as the length of the branches increases, the propagating branch gradually approaches the direction perpendicular to the applied

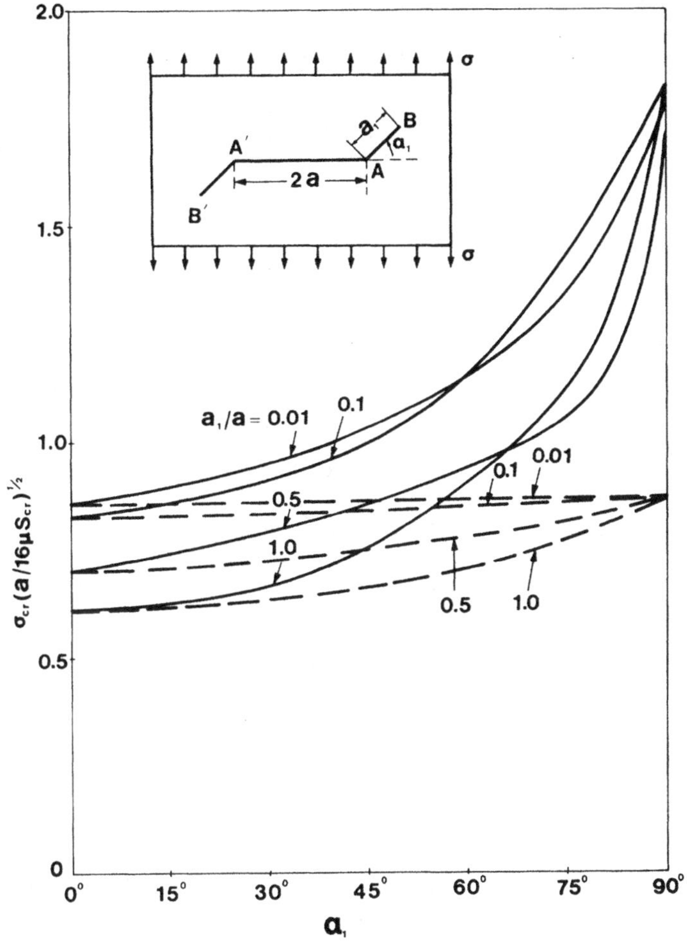

Figure 3.14. Variation of the quantity $\sigma_{cr}(a/16\mu S_{cr})^{1/2}$ versus the branch inclination angle α_1 for the skew-symmetric bent crack of Figure 3.13a for $a_1/a = 0.01$, 0.1, 0.5 and 1.0. Dotted curves indicate the values of the above quantity for the equivalent crack of length $2(a + a_1 \cos\alpha_1)$.

stress. For the case when $\alpha_1 = 90°$ the angle ϕ takes the values 56.7°, 72.4°, 90.2° and 95.8° for $a_1/a = 0.01, 0.1, 0.5$ and 1.0 respectively.

Let us now consider the skew-symmetric bent crack subjected to a tensile load inclined at an angle equal to 45° with respect to the plane of the main branch AA' (Figure 3.13b). The variation of the quantity $\sigma_{cr}(a/16\mu S_{cr})^{1/2}$ versus the angle α_1 of inclination of the branches for $a_1/a = 0.01, 0.1, 0.5$ and 1.0 is shown in Figure 3.16. It is observed that the minimum value of the critical stress happens for an angle α_1 close to 45°, which means that the branches are perpendicular to the direction of the applied stress. We also see that σ_{cr} decreases as the length a_1 of the branches increases. Defining the equivalent crack as in the previous case and calculating the corresponding critical stress, shown in Figure 3.16 by dotted lines, we find that the critical stress of

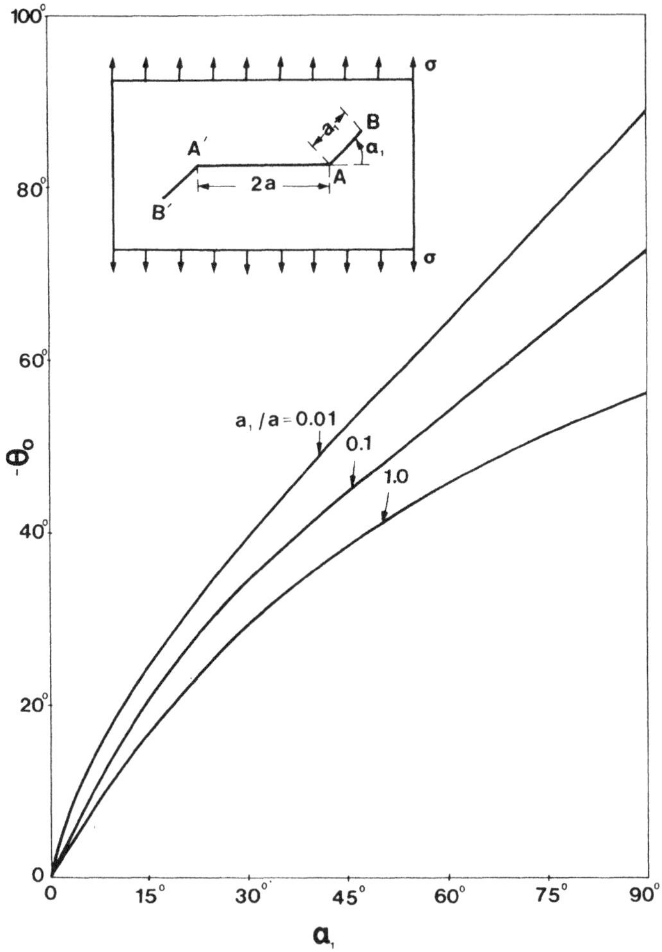

Figure 3.15. Values of the angle $-\theta_0$ that the fracture path originating from the tip *B* of the branch *AB* of the crack of Figure 3.13a forms with the line *AB*.

the bent crack is always smaller than that of the equivalent crack. If, however, we define the equivalent crack by the projection of the bent crack in the direction perpendicular to the applied stress σ this is not always the case. The values of this critical stress for this definition of the equivalent crack are shown in figure by dotted-with-period lines. It is thus concluded from these results that for the inclined bent crack, unlike to the cases of the doubly symmetric branched crack and the bent crack loaded perpendicularly to its plane, the critical stress is smaller than the corresponding stress for the equivalent crack for most geometrical configurations. This result can be attributed to the fact that for the case under consideration the inclined branches are closer to the direction perpendicular to the applied load than in the cases studied previously.

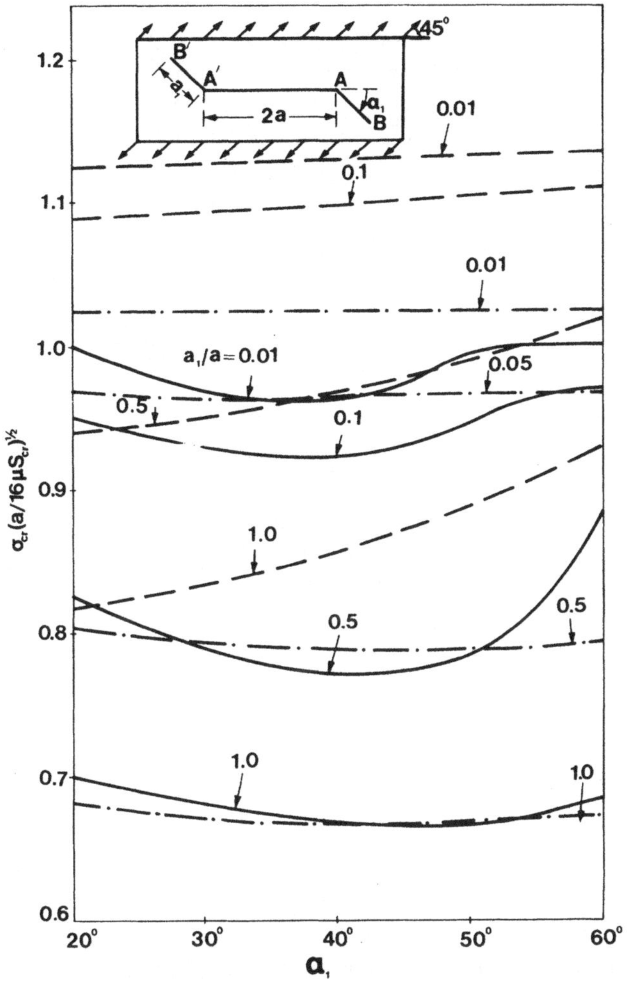

Figure 3.16. Variation of the quantity $\sigma_{cr}(a/16\mu S_{cr})^{1/2}$ versus the branch inclination angle α_1 for the skew-symmetric bent crack of Figure 3.13b for $a_1/a = 0.01, 0.1, 0.5$ and 1.0. Dotted curves indicate the values of the above quantity for a line straight crack of length $2(a + a_1\cos\alpha_1)$, while dotted-with-period lines stand for a line straight crack resulting from the projection of the crack to the direction perpendicular to the applied stress.

The values of the angle θ_0 that the propagating branch AB forms with its initial direction are shown in Figure 3.17. It is observed that the angle θ_0 may be positive or negative, and that it increases with the inclination angle α_1 of the branches. We also see that θ_0 decreases as the branch length a_1 increases. For those particular cases of the bent crack for which $\theta_0 = 0$ the branches extend in their own plane. These geometrical configurations of the crack are defined by: $a_1 = 0.01a$ and $\alpha_1 = 32.5°$; $a_1 = 0.1a$ and $\alpha_1 = 33.9°$; $a_1 = 0.5a$ and $\alpha_1 = 38.3°$; $a_1 = a$ and $\alpha_1 = 41.5°$.

b. *The bent crack with two branches (Figure 3.13c)*: Consider the bent crack of

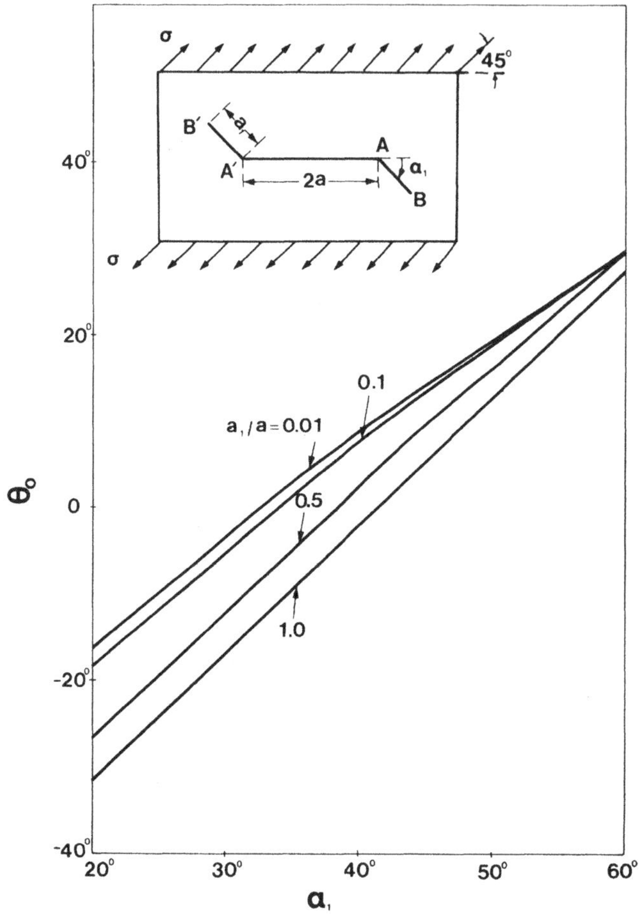

Figure 3.17. Values of the angle θ_0 that the fracture path starting from the tip B of the branch AB of the crack of Figure 3.13b forms with the line AB.

Figure 3.13c, composed of the main branch AC of unit length and the inclined branch CB of length a_1 which subtends an angle $-\alpha_1$ with AC. The crack is embedded in an infinite isotropic elastic body subjected to a uniform uniaxial stress perpendicular to the plane of the main branch AC. This type of crack branching has been considered by Chatterjee [7] who determined the values of k_1 and k_2 stress intensity factors at both crack tips A and B. Although the applied stress is perpendicular to the main crack AC, the stress field at the tip A of the branch AC is governed by both stress intensity factors k_1 and k_2 due to the existence of the inclined branch BC. Using the values of stress intensity factors in combination with the equations of the strain energy density criterion the critical stresses for possible extension of the bent crack from its tips A and B are determined. It is evident that the crack would extend from the tip with the minimum required stress for extension. For the geometrical configurations of the bent crack considered ($0 < a_1 < 1$ and $0 < \alpha_1 < 90°$) it was found that the crack always

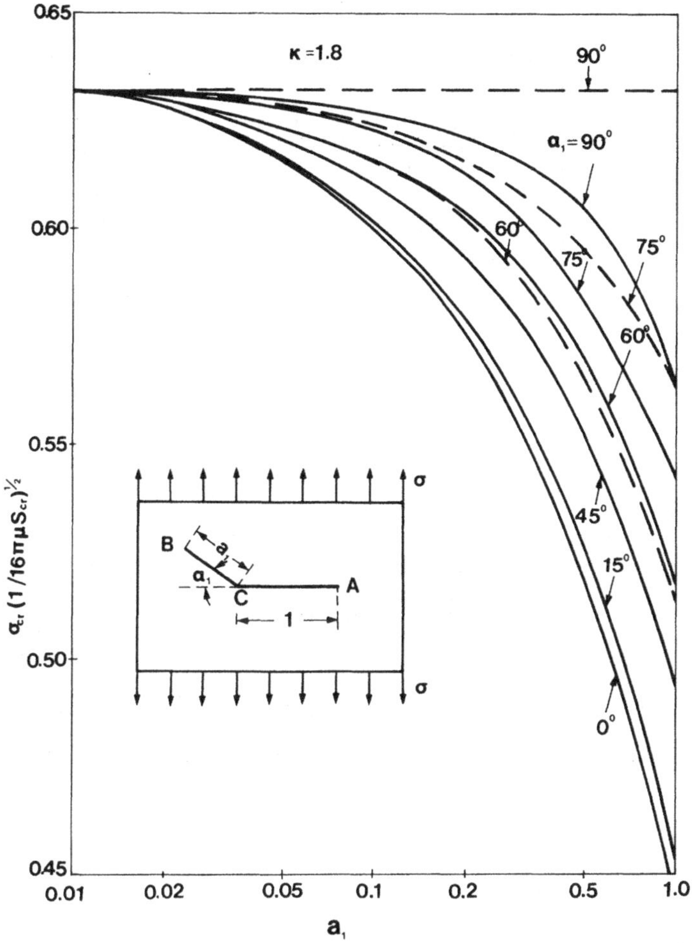

Figure 3.18. Variation of the quantity $\sigma_{cr}(1/16\pi\mu S_{cr})^{1/2}$ versus the branch crack length a_1 for the case of Figure 3.13c for the values of the inclination angle α_1 of the branch BC equal to 0°, 15°, 45°, 60°, 75° and 90°. Dotted curves indicate the values of the above quantity for the equivalent crack of length $(1 + a_1 \cos\alpha_1)$.

extends from tip A of the main branch AC. Figure 3.18 displays the variation of the normalized critical stress $\sigma_{cr}(1/16\pi\mu S_{cr})^{1/2}$ versus the length a_1 of the inclined branch for $\alpha_1 = 0°$, 15°, 45°, 60°, 75° and 90°. It is observed that for a given value of the angle of inclination α_1 of the inclined branch BC the critical stress decreases as its length a_1 increases. Furthermore, we observe that for a given length a_1 of the inclined branch the critical stress increases as the inclination α_1 increases. As in the cases of branched cracks we define the equivalent crack as that resulting from the projection of the inclined branch to the plane of the main branch (length of equivalent crack equal to $(1 + a_1 \cos\alpha_1)$). The critical values for brittle instability of the equivalent crack are presented in Figure 3.18 by dotted lines for $\alpha_1 = 0°$, 60°, 75° and 90°. It is observed that for $\alpha_1 = 60°$ the critical values of stress for propagating the equivalent crack are

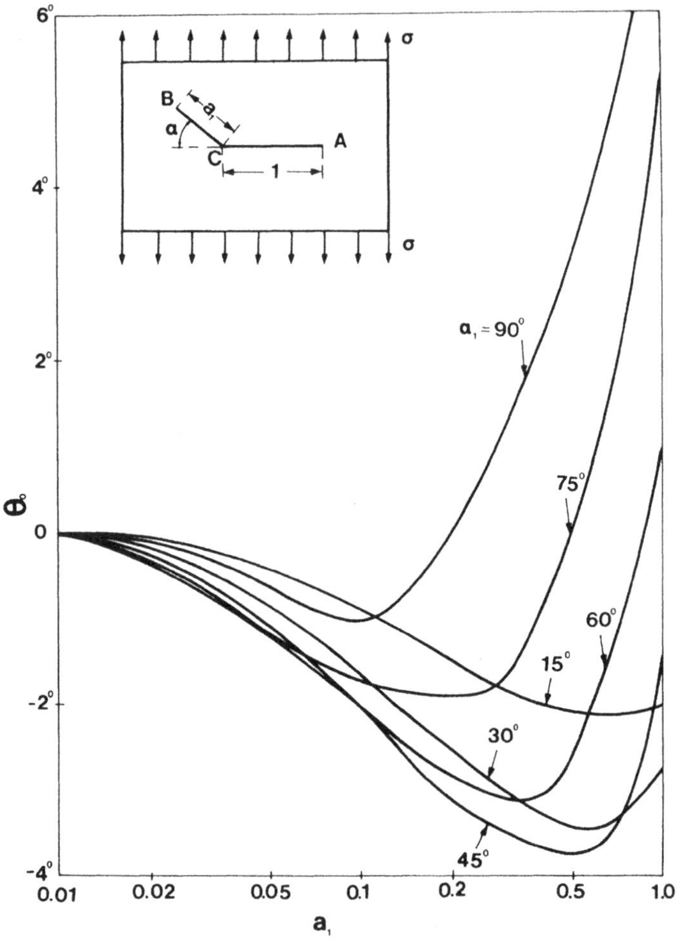

Figure 3.19. Values of the angle θ_0 that the fracture path starting from the tip A of the crack of Figure 3.13c forms with the line AC.

always smaller than those of the bent crack, while for $\alpha_1 = 75°$ and $90°$ this rule is reversed. We also see that the critical stress for extending the bent crack is always smaller than the stress required for extension of the main branch AC.

The values of the angle θ_0 that the fracture path originating from the tip A forms with the main branch AC are shown in Figure 3.19. Angle θ_0 is the polar angle in a Cartesian coordinate system centered at A, and the positive x-axis is oriented outward the branch AC. We observe that for branch angles α_1 smaller than $45°$ the angle θ_0 is always negative; that is the fracture path starting from A is clockwise relative to the main branch AC. However, for larger angles α_1 and large lengths of the inclined crack this rule is reversed. Thus the branch AC propagates in its own plane for $\alpha_1 = 60°$ and $a_1 = 0.9a$, for $\alpha_1 = 75°$ and $a_1 = 0.5a$ and for $\alpha_1 = 90°$ and $a_1 = 0.2a$. For values of a_1 greater than these limits for each angle α_1 the angle θ_0 is positive.

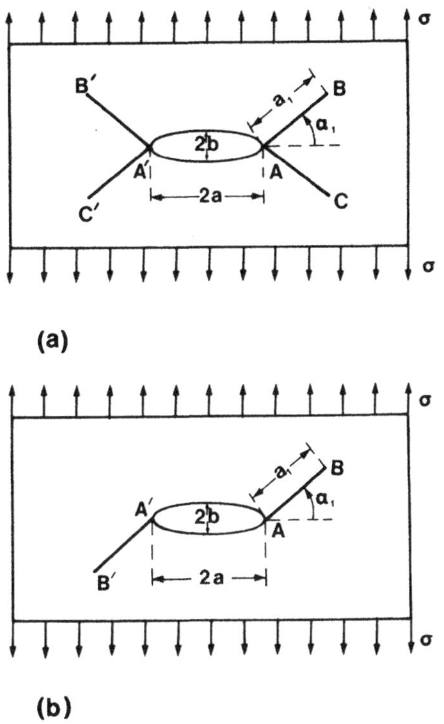

Figure 3. 20. Geometrical configurations of branched elliptical cracks.

3.5. Crack branches emanating from an elliptical crack

The branched cracks studied in this section are composed of a main elliptical crack and crack branches emanating from its boundary. Two particular types of cracks will be considered. In the first case, two equal and symmetrically located branches emanate from each end point of the elliptical crack (Figure 3.20a). In the second case, one branch emanates from each end point of the elliptical crack with both these branches having the same length, anti-symmetrically located with respect to the elliptical crack (Figure 3.20b). These geometrical configurations are analogous to the doubly symmetric branched crack of Figure 3.1a and the skew-symmetric branched crack of Figure 3.13a. The only difference noted is that the main branch is elliptical rather than sharp. Therefore, the objective of this section is to study the influence of the root radius of the main elliptical branch on the brittle fracture behavior of the branched crack. The doubly symmetric elliptical branched crack and the skew-symmetric elliptical branched crack will be considered individually.

a. *The doubly symmetric elliptical branched crack (Figure 3.20a)*: Consider the crack of Figure 3.20a embedded in an infinite isotropic elastic plate subjected to a uniform uniaxial stress σ at infinity perpendicular to the axis of symmetry of the elliptical crack. For this case with a crack root radius $\rho/a = (b/a)^2 = 0.5$ in a plate

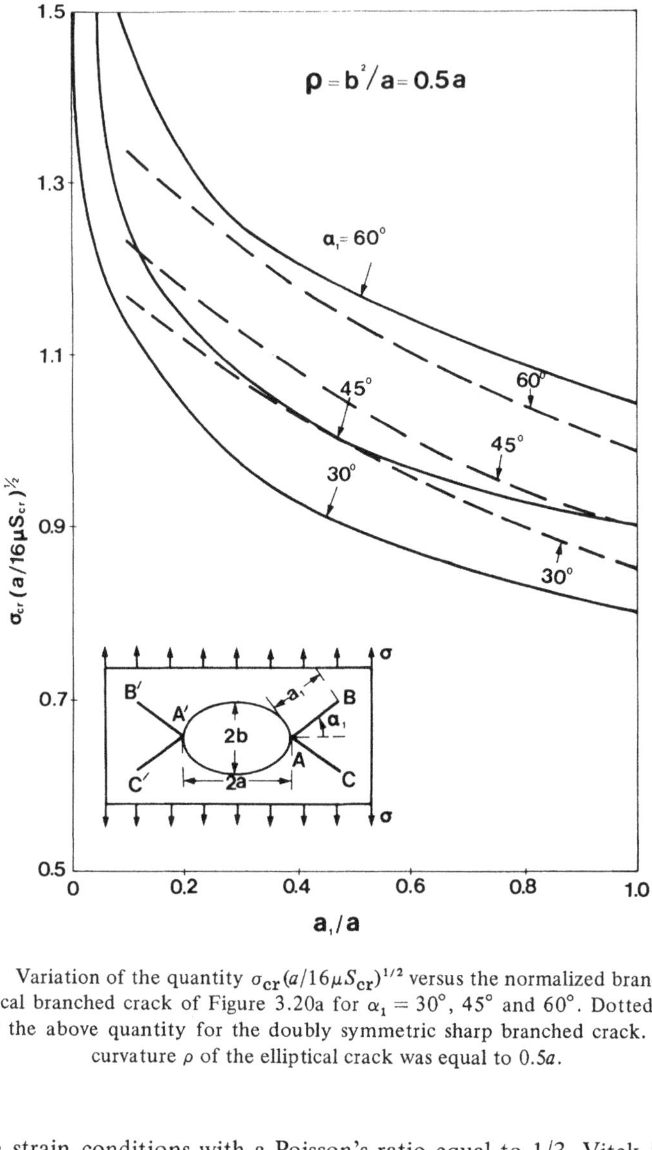

Figure 3.21. Variation of the quantity $\sigma_{cr}(a/16\mu S_{cr})^{1/2}$ versus the normalized branch length a_1/a for the elliptical branched crack of Figure 3.20a for $\alpha_1 = 30°$, $45°$ and $60°$. Dotted lines indicate the values of the above quantity for the doubly symmetric sharp branched crack. The radius of curvature ρ of the elliptical crack was equal to $0.5a$.

under plane strain conditions with a Poisson's ratio equal to 1/3, Vitek [9] gave the values of k_1 and k_2 stress intensity factors at the tips of the branches. Using the equations of the strain energy density criterion the critical value σ_{cr} of the applied stress for crack extension is determined as a function of the length and inclination of the branches. Figure 3.21 presents the variation of the quantity $\sigma_{cr}(a/16\mu S_{cr})^{1/2}$ versus the length a_1 of the branches normalized to the half length of the major axis of the elliptical branch a for $\alpha_1 = 30°$, $45°$ and $60°$. As in the case of the doubly symmetric branched crack, the critical stress decreases as either the length of the branches increases or their angle of inclination decreases. In order to compare the cases of the elliptical and sharp branched cracks, the values of the critical stress for the sharp

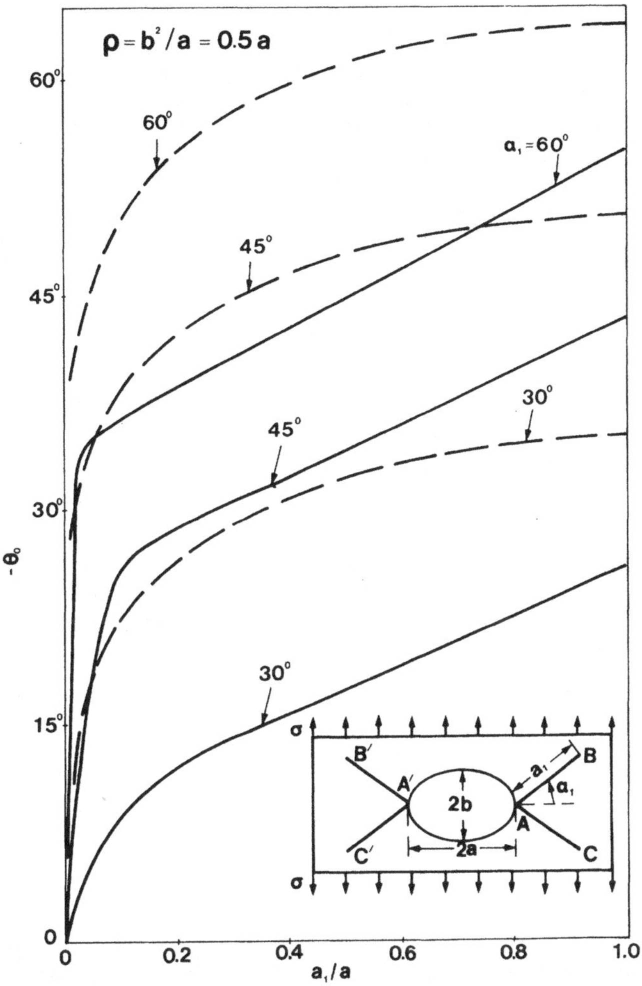

Figure 3.22. Values of the angle $-\theta_0$ that the fracture path starting from the tip B of the branch AB of the crack of Figure 3.20a forms with the line AB. Dotted lines indicate the values of the above quantity for the doubly symmetric sharp branched crack. The radius of curvature ρ of the elliptical crack was equal to $0.5a$.

branched crack are shown in Figure 3.21 by dotted lines for the same angles of inclination of the branches. It is observed that for $\alpha_1 = 30°$ and $45°$ the critical stress for the elliptical crack is smaller than the corresponding stress for the sharp crack, while for $\alpha_1 = 60°$ this rule is reversed. Thus, it is concluded that for large branch angles the existence of a radius of curvature at the root of the main branch increases the strength of the cracked plate, while the contrary occurs for small branch angles.

The values of the angle formed by the fracture path starting from the tip B of the branch AB with the line AB are shown in Figure 3.22. As in the case of the doubly symmetric branched crack for the branch angles of Figure 3.22 the angle θ_0 is negative,

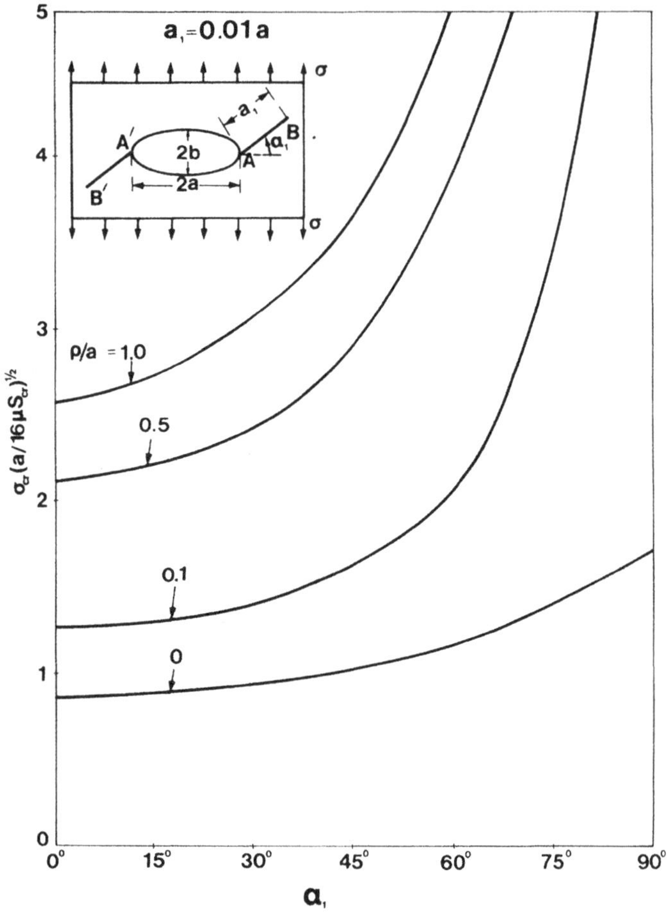

Figure 3.23. Variation of the quantity $\sigma_{cr}(a/16\mu S_{cr})^{1/2}$ versus the angle α_1 of inclination of the branches for the crack of Figure 3.20b for $\rho/a = 0$, 0.1, 0.5 and 1.0. The value $\rho = 0$ corresponds to the sharp branched crack. The length a_1 of the branches is equal to $0.01a$.

that is, branches tend to propagate with decreasing angle, as if the branches attract each other. In order to compare the case of the doubly symmetric branched crack, the values of the angle θ_0 for this case are displayed in Figure 3.22 by dotted lines. It is observed that the angles θ_0 for the elliptical crack are always smaller than for the sharp crack.

b. *The skew-symmetric elliptical branched crack (Figure 3.20b)*: The critical fracture stress and the fracture angle are determined using the values of k_1 and k_2 from [9], as in the previous case. Figures 3.23 and 3.24 present the variation of the quantity $\sigma_{cr}(a/16\mu S_{cr})^{1/2}$ versus the angle α_1 of inclination of the branches for the values of the root radius of the elliptical crack equal to $\rho/a = 0$, 0.1, 0.5 and 1.0 when the branch length a_1 takes the value $0.01a$ and $0.5a$ respectively. The value of $\rho = 0$ corresponds to the sharp bent crack studied in Section 3.4a, while the value $\rho = a$ corresponds to

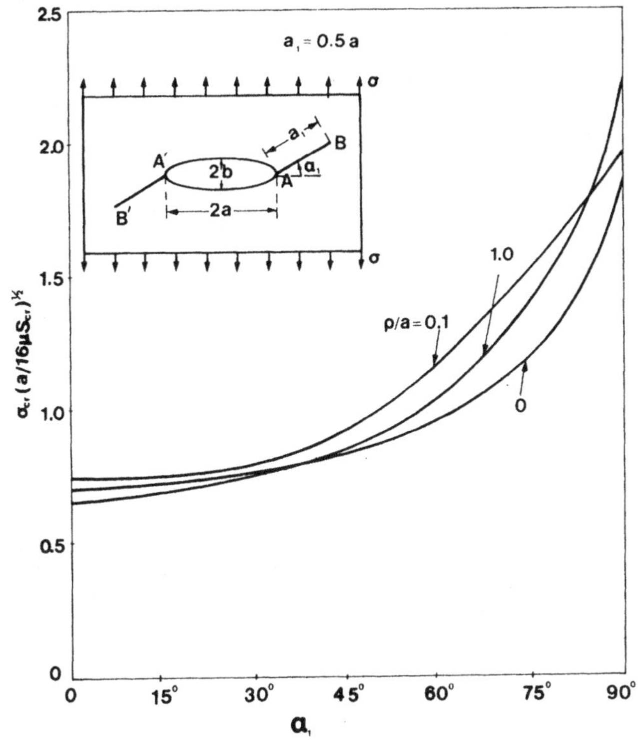

Figure 3.24. As in Figure 3.23 for $a_1 = 0.5a$.

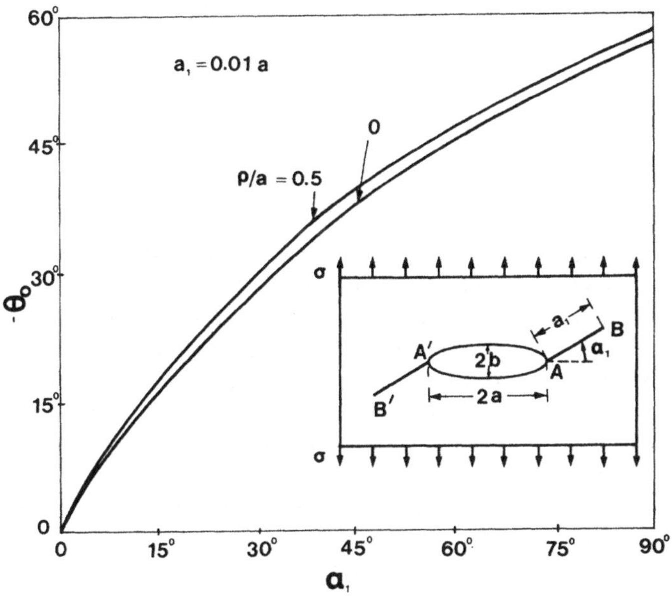

Figure 3.25. Values of the angle $-\theta_0$ that the fracture path originating from the tip B of the crack of Figure 3.20b forms with the line AB. The length a_1 of the branches is equal to $0.01a$.

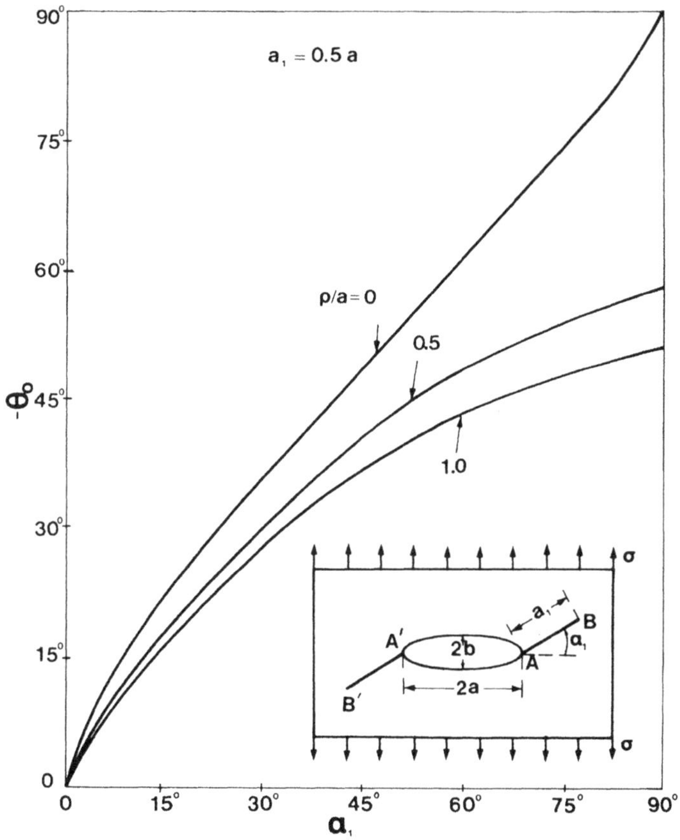

Figure 3.26. As in Figure 3.25 for $a_1 = 0.5a$.

the case of a circular hole. The effect of the root radius of the main branch on the critical stress can be established from these figures. It is observed that for small branch lengths ($a_1 = 0.01a$) the critical stress increases as the root radius ρ increases. Therefore, the most dangerously shaped main crack on the strength of the body is the sharp crack. However, when the branch length a_1 increases no such rule can be stated. As in the case of the sharp bent crack, the critical stress σ_{cr} increases with the inclination of the branches for all root radii of the elliptical crack. Furthermore, the previously established effect of the decrease of the critical stress with the increase of the branch length is also observed.

The fracture angles $-\theta_0$ for the branch AB for the cases of Figures 3.23 and 3.24 are displayed in Figures 3.25 and 3.26 respectively. We see that, as in the case of the bent crack, the fracture angle is negative. This means that the branches have a tendency to propagate in a direction perpendicular to the applied stress. The angle $-\theta_0$ increases with the angle of inclination α_1 of the branches.

References

[1] Andersson, H., Stress-intensity factors at the tips of a star-shaped contour in an infinite tensile sheet, *Journal of the Mechanics and Physics of Solids*, 17, pp. 405–417 (1969).

[2] Andersson, H., Erratum to [1], *Journal of the Mechanics and Physics of Solids*, 18, p. 437 (1970).

[3] Theocaris, P.S., Complex stress-intensity factors at bifurcated cracks, *Journal of the Mechanics and Physics of Solids*, 20, pp. 265–279 (1972).

[4] Theocaris, P.S. and Blonzou, C.H., Symmetric branching of cracks in PMMA (Plexiglas), *Materialprufüng*, 15, pp. 123–130 (1973).

[5] Theocaris, P.S. and Ioakimidis, N., The symmetrically branched crack in an infinite elastic medium, *Zietschrift fur angewandte Mathematik und Physik (ZAMP)*, 27, pp. 801–814 (1976).

[6] Theocaris, P.S., Asymmetric branching of cracks, *Journal of Applied Mechanics*, 44, pp. 611–618 (1977).

[7] Chatterjee, S.N., The stress field in the neighborhood of a branched crack in an infinite elastic sheet, *International Journal of Solids and Structures*, 11, pp. 521–538 (1975).

[8] Kitagawa, H., Yuuki, R. and Ohira, T., Crack morphological aspects in fracture mechanics, *Engineering Fracture Mechanics*, 7, pp. 515–529 (1975).

[9] Vitek, V., Plane strain stress intensity factors for branched cracks, *International Journal of Fracture*, 13, pp. 481–501 (1977).

[10] Kitagawa, H. and Yuuki, R., Analysis of branched cracks under biaxial stresses, *Advances in Research on the Strength and Fracture of Materials*, Vol. 3A – Analysis and Mechanics, Editor D.M.R. Taplin, Pergamon Press, pp. 201–211 (1978).

[11] Kalthoff, J.F., On the characteristic angle for crack branching in brittle materials, *International Journal of Fracture Mechanics*, 7, pp. 478–480 (1971).

4

Interacting cracks

4.1. Introduction

It has been established that the mechanism of fracture of a material in the microscale is closely associated to the interaction, coalescence, growth and propagation of minute defects such as cracks, voids, inclusions, etc. Consequently, the resistance of a material to fracture under any form of loading depends on the conditions in which the aforesaid defects interact, coalesce, grow and propagate. The study of the interaction of the above defects existing in a body subjected to a given form of loading is therefore of particular importance.

Interaction effects for cracks lying on the same line and for other more complicated geometrical configurations have been studied in references [1–15] and [16–27] respectively. In these cases attention was focused mainly on the determination of the opening-mode and sliding-mode stress intensity factors at the interacting crack tips.

The objective of this chapter is to study the brittle fracture characteristic phenomena which take place in isotropic homogeneous elastic plates with interacting cracks of various geometrical configurations. The cases of two equal skew-parallel cracks, two equal symmetrically or arbitrarily oriented cracks and X-formed arrays of cracks will be studied in detail. In all these cases the critical values of the applied loads for crack extension are calculated. Furthermore, the more vulnerable crack tip from which fracture originates and the fracture path are determined. Important results establishing the dependence of the critical stress of fracture and the pattern of initial crack propagation on the geometrical configuration of the existing cracks and the loading of the plate are disclosed by this study.

4.2. Two equal skew-parallel cracks

Consider an infinite isotropic elastic plate containing two skew-parallel cracks, each of length $2a$, and subjected to a uniform uniaxial remote stress σ perpendicular to the axis of the cracks (Figure 4.1). The distance between the outer tips of the cracks measured parallel to their direction is $2b$ and the perpendicular distance between the cracks is $2h$. This configuration has been studied by Yokobori *et al.* [17] and Isida [18] who gave the values of both k_1 and k_2 stress intensity factors for crack tips A and B. Their results have been reported by Sih [26] and Rooke and Cartwright [27].

The critical stress $\sigma_{cr}^{A,B}$ required for possible crack extension from tips A and B is determined by introducing the values of k_1 and k_2 into the equations of the strain energy density criterion for the general two-dimensional crack problem (Section 2.2, page 11). It is evident that the actual critical stress σ_{cr} of the plate for fracture

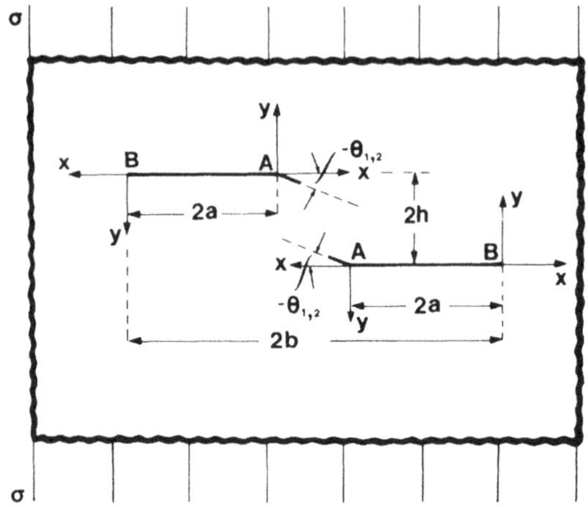

Figure 4.1. Geometry of a cracked plate containing two equal skew-parallel cracks.

instability is equal to the minimum value of the critical stresses σ_{cr}^A and σ_{cr}^B required for crack initiation from the tips A and B. This fact allows the determination of the more vulnerable crack tip from which extension of the crack AB starts, as well as the initial crack propagation pattern. The cases when the applied stress σ is either tensile or compressive are considered.

Figures 4.2a, 4.2b and 4.3 present the variation of the quantity $\sigma_{1cr}(a/16\mu S_{cr})^{1/2}$ versus a/b for $h/a=0.1$, 0.5 and 1.0 respectively and tensile applied stress σ_1. The regions in which crack extension starts either from the tip A or the tip B are separated by dotted lines, and are indicated by the letters A and B in the corresponding areas of the figures. Propagation of the interacting cracks starts from their inner tips for $h/a=0.1$ and $0<a/b<0.515$, for $h/a=0.5$ and $0<a/b<0.505$ and for $h/a=1.0$ and $0<a/b<0.405$. Thus, for large horizontal distances between the two cracks, crack propagation starts from the inner tips, while when the two cracks approach each other, crack propagation takes place from the outer tips, depending on the value of h/a. The value of the horizontal distance b/a separating the outer tips for which the inner tips propagate increases as the perpendicular distance h/a increases. These results are in agreement with those found previously by Lange [16] who studied the interaction between two skew-parallel cracks in a tensile stress field for the case when a variable overlap exists. Lange determined photoelastically the principal stress trajectories and the curves of constant maximum shear stress for three different geometrical configurations of the parallel cracks corresponding to zero and a smaller or greater overlap distance than the perpendicular distance of the cracks. He found that once overlap occurs the far crack tips exhibit higher order isochromatics than at the inner tips. This means that the stress magnitude at the far tips is larger than the inner tips. Therefore, crack propagation starts from the far tips. Lange was unable to make a thorough study of the phenomenon by studying only three different geometries of the cracks, so his results are only of qualitative character, However, they provide a good indication of the overlap effect and compare favorably with our results.

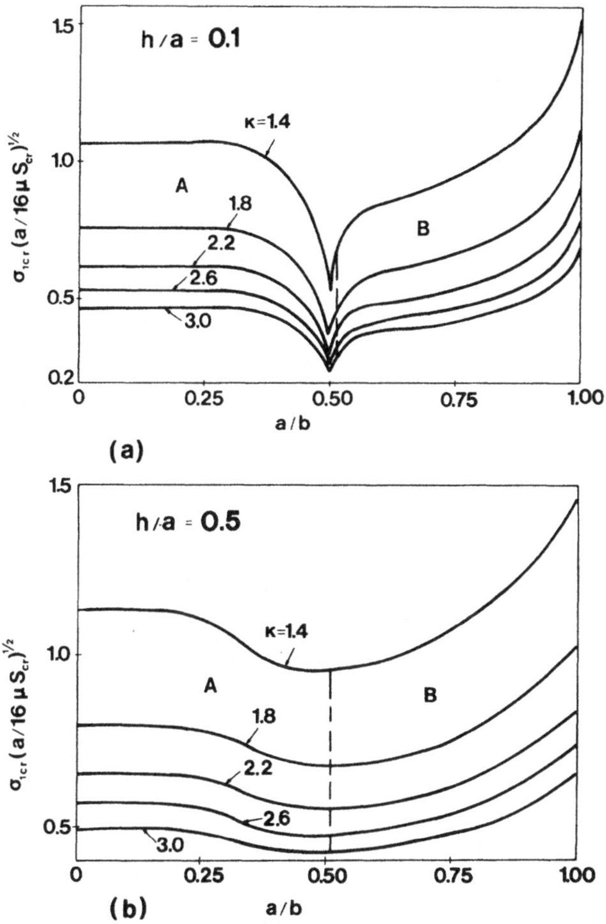

Figure 4.2. Variation of the normalized critical stress σ_{1cr} for crack propagation of the plate with two equal skew-parallel cracks versus a/b for $h/a = 0.1$ (a) and $h/a = 0.5$ (b). The cracked plate is subjected to tensile loads and κ takes the values: 1.4, 1.8, 2.2, 2.6 and 3.0.

From Figures 4.2 and 4.3 it can be deduced that for each value of the perpendicular distance h/a of the cracks there exists a definite value of their horizontal distance b/a for which the stress σ_{1cr} becomes minimum. This critical distance b/a increases as the perpendicular distance of the cracks increases. Thus σ_{1crmin} occurs at $b/a = 2.0, 2.22, 2.50$ and 2.63 for $h/a = 0.1, 0.5, 0.8$ and 1.0 respectively. We observe that all these values of b/a are very close and, in any case, are greater than the limiting values of b/a for which transition of crack propagation from the inner to the far tips takes place. It is therefore concluded that the minimum value of the critical stress always corresponds to crack propagation from the inner tips.

Figures 4.2 and 4.3 also show that the critical stress σ_{1cr} for crack propagation increases for each particular value of the horizontal distance b/a between the cracks as the perpendicular distance h/a increases. Table 4.1 shows the normalized minimum

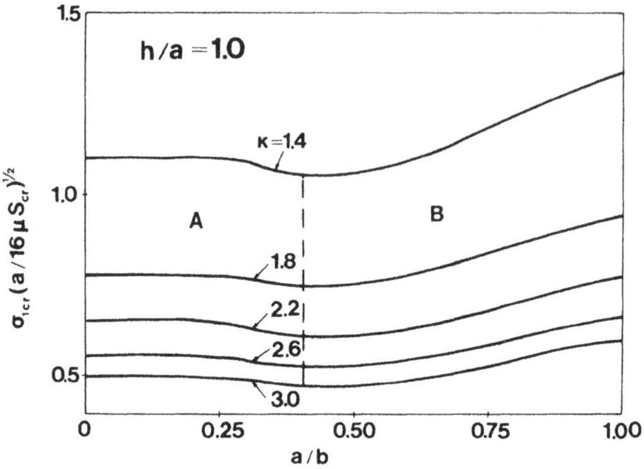

Figure 4.3. As in Figure 4.2 for $h/a = 1.0$.

Table 4.1. Minimum values of $\sigma_{1\text{cr min}}$ of the critical stress for crack propagation of a plate with two equal skew-parallel cracks loaded in tension and the corresponding critical stresses $\sigma'_{10\text{cr}}$ and $\sigma_{10\text{cr}}$ for a single crack of length equal to the length of each separate crack or to the combined length of the two cracks. All these values are normalized to $(16\mu S_{\text{cr}}/a)^{1/2}$.

h/a	$\kappa = 1.4$	$\kappa = 1.8$	$\kappa = 2.2$	$\kappa = 2.6$	$\kappa = 3.0$
			$\sigma_{1\text{cr min}}$		
0.1	0.53	0.37	0.30	0.26	0.24
0.5	0.94	0.66	0.54	0.47	0.42
0.8	1.02	0.72	0.59	0.51	0.46
1.0	1.06	0.75	0.61	0.53	0.47
			$\sigma'_{10\text{cr}}$		
	1.12	0.79	0.65	0.56	0.50
			$\sigma_{10\text{cr}}$		
0.1	0.79	0.56	0.46	0.40	0.36
0.5	0.75	0.53	0.44	0.38	0.34
0.8	0.71	0.50	0.41	0.35	0.32
1.0	0.69	0.49	0.40	0.34	0.31

values $\sigma_{1\text{cr min}}$ of $\sigma_{1\text{cr}}$ for various values of h/a. We observe that for each particular value of κ, $\sigma_{1\text{cr min}}$ increases as h/a also increases.

It is now of particular importance to compare the above minimum values, $\sigma_{1\text{cr min}}$, of the critical stress, $\sigma_{1\text{cr}}$, with the corresponding values of this stress for a single crack of length equal either to the length of each separate crack (corresponding to the case when each crack acts separately from the other) or to the combined length of the two cracks defined by the distance between their outer tips (corresponding to the case when the two cracks act as one). The critical stresses $\sigma'_{10\text{cr}}$ and $\sigma_{10\text{cr}}$ for the first and second cases, respectively are shown in Table 4.1.

From a comparison of the values $\sigma_{1\mathrm{cr\,min}}$ with the corresponding values $\sigma'_{10\mathrm{cr}}$ and $\sigma_{10\mathrm{cr}}$ we conclude that, for small perpendicular distances between the two cracks, the minimum critical stress is less than the corresponding stress when the two cracks act as one. It is therefore concluded that for such cases, it is more dangerous for the plate to have two interacting cracks than one crack whose length is equal to the total length of the two cracks. Thus, for $h/a = 0.1$ the stress $\sigma_{1\mathrm{cr\,min}}$ is 50% smaller than the corresponding stress when the two cracks act as one. As the perpendicular distance h/a between the cracks increases, $\sigma_{1\mathrm{cr\,min}}$ approaches the value of the critical stress when the two cracks act as one, and for greater values of h/a, $\sigma_{1\mathrm{cr\,min}}$ becomes larger than this stress. Thus, for $h/a = 0.5$, $\sigma_{1\mathrm{cr\,min}}$ is 20% greater than $\sigma_{10\mathrm{cr}}$ and 20% smaller than $\sigma'_{10\mathrm{cr}}$. As the perpendicular distance h/a between the two cracks further increases the interaction effects become less and less significant and each of the two cracks acts as if the other were absent. Thus, for $h/a = 0.8$, the minimum critical stress $\sigma_{1\mathrm{cr\,min}}$ for failure of the plate with the two equal skew-parallel cracks is 10% smaller than the corresponding stress for failure of a plate with one such crack, while for $h/a = 1.0$ the above difference is only 6%.

The above comparison was made for each h/a for the minimum critical stress corresponding to a particular value of b/a as well as for other values of b/a. It was found that for $h/a = 0.1$, $\sigma_{1\mathrm{cr}}$ is equal to $\sigma_{10\mathrm{cr}}$ when $a/b = 0.47$ and 0.52. For the intermediate values of a/b $(0.47 < a/b < 0.52)$, $\sigma_{1\mathrm{cr}}$ is less than $\sigma_{10\mathrm{cr}}$, while for the interval $0.52 < a/b < 0.80$, $\sigma_{1\mathrm{cr}}$ is very close to $\sigma_{10\mathrm{cr}}$. As h/a increases, $\sigma_{1\mathrm{cr}}$ becomes progressively greater than $\sigma_{10\mathrm{cr}}$ for the whole range of values of a/b. Thus, for $h/a = 0.5$, $\sigma_{1\mathrm{cr}}$ is greater than $\sigma_{10\mathrm{cr}}$ by approximately 20% for $0.44 < a/b < 0.52$, and by 15% for $0.54 < a/b < 0.90$.

We are now able to compare the fracture stress $\sigma_{1\mathrm{cr}}$ of a plate with two equal parallel cracks which are not displaced along their direction $(a/b = 1.0)$ with the corresponding stress of a plate with one such crack. Table 4.2 shows the values of $\sigma_{1\mathrm{cr}}$ for various values of h/a. The value as $h/a \to \infty$ corresponds to the case when the plate contains one crack. It can be concluded from this table that the fracture stress, $\sigma_{1\mathrm{cr}}$, of the plate with two cracks is always greater than the same stress of the plate with one crack, and that $\sigma_{1\mathrm{cr}}$ increases as the perpendicular distance h/a between the cracks decreases.

The variation of the crack extension angle θ_1 versus a/b for both tips A and B and tensile applied loads is presented in Figures 4.4 and 4.5 for $h/a = 0.1$ and 1.0 respectively. In all these figures, vertical lines correspond to the critical values of the ratio a/b for which transition of crack propagation from tips A to tips B takes place.

Table 4.2. Fracture stresses, $\sigma_{1\mathrm{cr}}$, of a plate with two equal parallel cracks which are not displaced. All these values are normalized to $(16\mu S_{\mathrm{cr}}/a)^{1/2}$.

h/a	$\kappa = 1.4$	$\kappa = 1.8$	$\kappa = 2.2$	$\kappa = 2.6$	$\kappa = 3.0$
			$\sigma_{1\mathrm{cr}}$		
0.1	1.57	1.12	0.92	0.80	0.71
0.5	1.46	1.03	0.84	0.73	0.65
1.0	1.34	0.95	0.77	0.67	0.60
∞	1.12	0.79	0.65	0.56	0.50

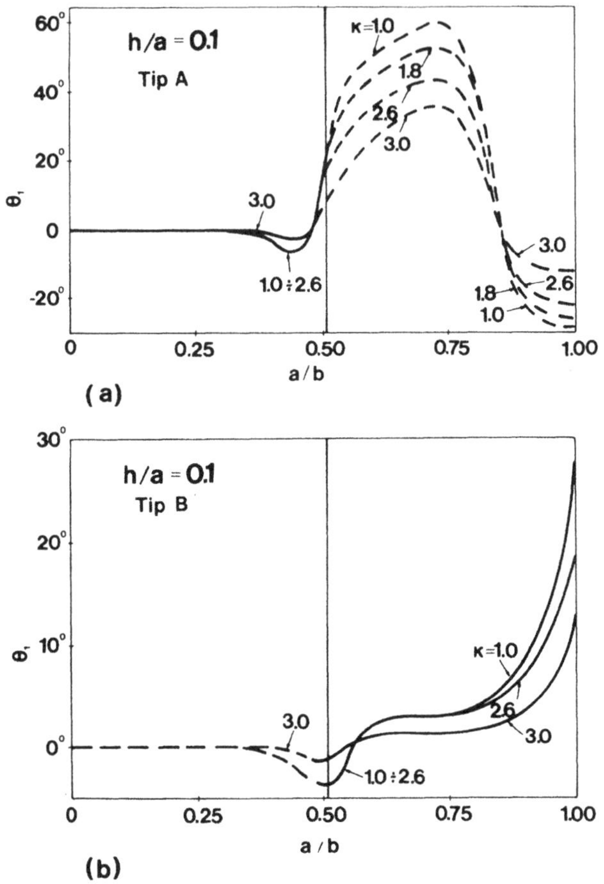

Figure 4.4. Variation of the angle θ_1 of initial crack propagation versus a/b for the crack tips A (a) and B (b) for $h/a = 0.1$. The cracked plate is subjected to tensile loads and κ takes the values: 1.0, 1.8, 2.6 and 3.0. The vertical lines of figure correspond to the limiting values of a/b for which transition of crack propagation from tip A to tip B takes place. All curves corresponding to actual crack propagation angles are shown by continuous lines.

These critical values of a/b, are 0.515, and 0.405 for $h/a = 0.1$ and 1.0 respectively, as indicated above. For all values of the ratio a/b smaller than these critical values, crack propagation occurs from tips A, while for the remaining values of a/b crack propagation starts from tips B. In Figures 4.4 and 4.5 all curves corresponding to the actual crack propagation angles from either crack tip A or B are presented by continuous lines, while the remaining curves are presented by dotted lines.

Figures 4.4 and 4.5 show that for large horizontal distances between the two cracks, propagation takes place from the inner tips of the cracks, while as the two cracks approach, the crack propagation starts from the outer tips. For the first case (of crack propagation from the inner tips), the intervals of a/b for $h/a = 0.1, 0.5$, and 1.0, for which the angle θ_1 becomes negative or positive are shown in Table 4.3. In the same table the maximum values, θ_{1max}, of θ_1 are also presented. Negative values

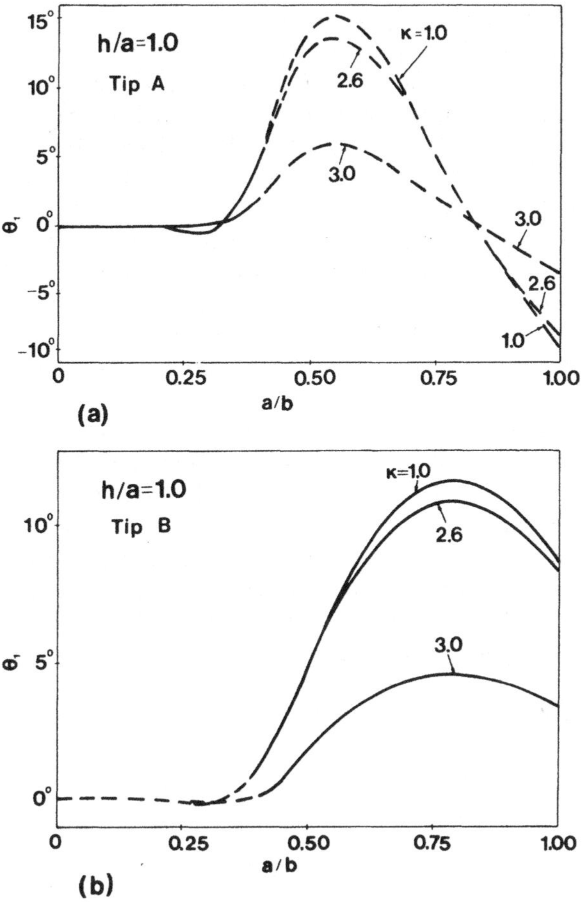

Figure 4.5. As in Figure 4.4 for $h/a = 1.0$.

Table 4.3. Intervals of a/b for which the angle θ_1 of crack propagation becomes negative or positive and values of a/b for which this angle becomes maximum. All values of this table correspond to crack propagation from the inner crack tip.

$h/a = 0.1$	$h/a = 0.5$	$h/a = 1.0$
	$\theta_1 < 0$	
(0, 0.481)	(0, 0.400)	(0, 0.320)
	$\theta_1 > 0$	
(0.481, 0.515)	(0.400, 0.505)	(0.320, 0.405)
	$\theta_{1\max}$	
0.440	0.505	0.405

of θ_1 correspond to a crack propagation pattern where one inner crack tip extends so as to join the other, while for positive values of θ_1 the one inner tip propagates in the opposite direction of the other so that the two cracks repel one another. For the second case, (of crack propagation from the outer tips), angle θ_1 is always positive (with the exception of the values of a/b lying in the interval $0.515 < a/b < 0.565$ for $h/a = 0.1$). This means that the cracks propagate in opposite directions.

Theocaris established experimentally the fact that once overlap occurs, the two cracks curve towards one another [28]. He determined the propagation paths of two unequal edge cracks subjected to quasi-static or dynamic loads by the optical method of caustics in combination with a high speed camera, and established interesting results for the crack propagation velocities and the crack tip stress intensity factors. For the case when crack propagation takes place under quasi-static loading, Theocaris found that only the longer crack propagates, while the shorter crack remains stationary. Since one crack is many times longer than the other when overlap occurs, the experimental results of this case cannot be compared with ours, which are referred to cracks of equal length. However, for dynamic loading, and under certain favorable geometrical configurations of the cracks, both edge cracks propagate simultaneously, and when they overlap their lengths are almost equal. For such a case Theocaris found that the two cracks deviate towards one another (see Figure 5, reference [28] corresponding to the time interval $144\,\mu\text{sec} < t < 352\,\mu\text{sec}$, where time, t, is measured from the beginning of crack propagation), which agrees with our results.

Figures 4.6 and 4.7 present the variation of the critical stress σ_{2cr} and the crack extension angle θ_2 from either tip A or B versus a/b, respectively, for the case when the applied stress σ is compressive and $h/a = 0.5$. Observations analogous to the case of tensile loads can be made from these figures and the corresponding ones for other values of h/a.

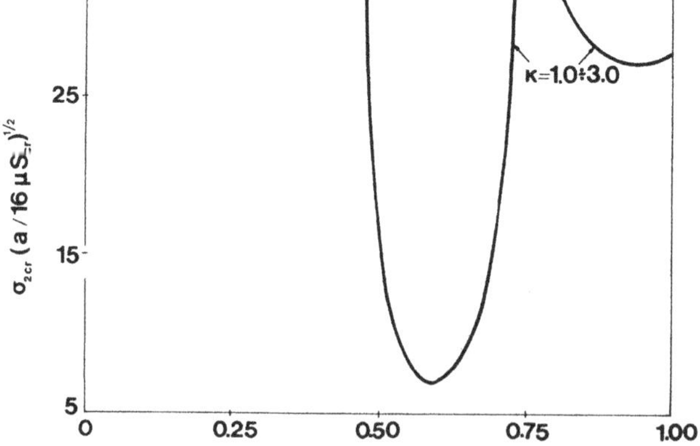

Figure 4.6. As in Figure 4.2b for compressive applied loads.

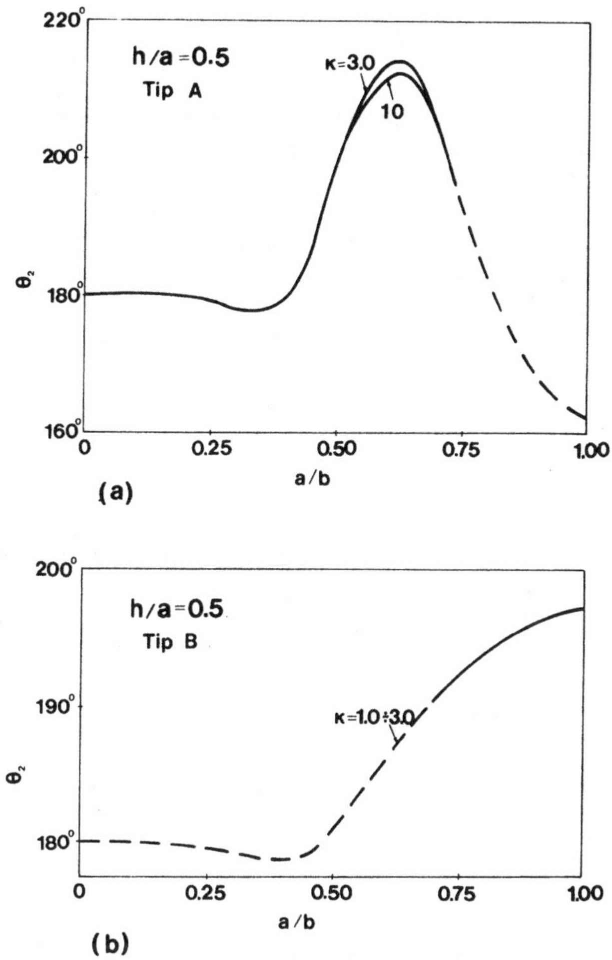

Figure 4.7. As in Figure 4.4 for compressive applied loads and $h/a = 0.5$.

The basic results of this section concerning the dependence of the critical fracture stress and the initial crack propagation pattern on the geometrical configuration of the two interacting cracks may be summarized as follows:

(i) Crack propagation starts from the inner or far tips of the interacting cracks, depending on their particular geometrical configuration. Where a great horizontal distance exists between the two cracks, crack propagation starts from the inner tips. As the two cracks approach each other, however, the propagation starts from the far tips. The limits of the horizontal distances of the cracks at which the one or the other situation takes place were established above as a function of the perpendicular distance of the cracks.

(ii) For each particular value of the perpendicular distance of the cracks there is a critical value of their horizontal distance for which the critical stress becomes minimum. This distance which corresponds always to crack propagation from the inner tips as well as the minimum stress increase as the perpendicular distance of the cracks increases. Numerical results of the critical distances and the corresponding minimum stresses were given above.

(iii) A comparison of the critical stress for crack propagation with the corresponding stress when the two cracks act either as one or separately from one another was made. The limits of the distances of the two cracks for which the one or the other situation occurs were established. Further, for small perpendicular distances between the two cracks, it was found that the critical stress is less than the corresponding stress when the two cracks act as one.

(iv) The fracture load of a plate with two equal parallel cracks not displaced along their direction is always greater than the corresponding load for one such crack. It increases as the perpendicular distance of the cracks decreases.

(v) The fracture load of the plate increases as its Poisson's ratio increases.

(vi) For the geometrical configurations in which crack propagation starts from the inner tips there is a limiting value of the horizontal distance of the cracks at which they propagate in their own plane. For greater values of the horizontal distance, one inner tip propagates in the direction of the other, while for smaller distances both propagate in opposite directions. However, the two cracks always propagate in opposite directions when the propagation begins from the far tips.

For further information on this study the interested reader is referred to reference [29].

4.3. Two equal symmetrically inclined cracks

Consider the two cracks AB and $A'B'$ of Figure 4.8, each of length $2a$ and equally inclined under angles $(\phi/2)$ to the vertical direction. The horizontal distance between the centers of the cracks is equal to $2.2a$. Both cracks are embedded in an infinite isotropic elastic plate subjected to a uniform uniaxial tensile stress in either the vertical

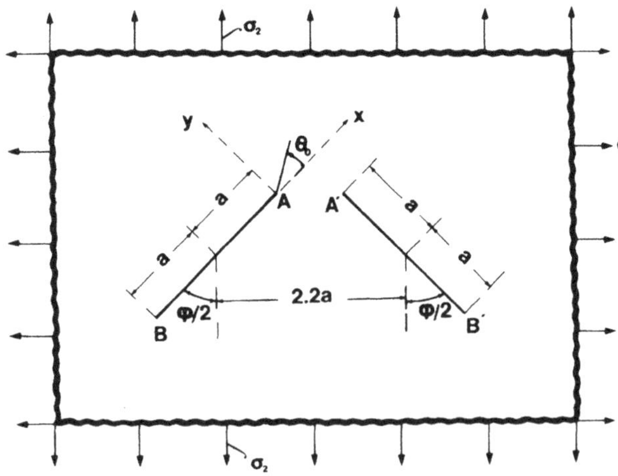

Figure 4.8. Geometry of a cracked plate with two equal and equally inclined to the direction of the applied stress cracks.

or the horizontal direction. This situation has been considered by Aksogan [21], who determined the values of k_1 and k_2 stress intensity factors for both tips A and B of the crack AB under the applied stress σ_1 or σ_2. In this section the brittle fracture behavior of the plate will be studied based on these values of stress intensity factors in combination with the strain energy density criterion. The cases when the plate is subjected either to the stress σ_1 or to the stress σ_2 will be considered individually.

a. *The stress is applied along the vertical direction.* Let σ_2 be the applied stress and $(\phi/2)$ the angle that each crack subtends with the direction of σ_2. Introducing the values of k_1 and k_2 for both tips A and B of the crack AB into the equations of the strain energy density criterion (Section 2.2, page 11) the critical stresses σ_{cr}^A and σ_{cr}^B for possible extension of the crack AB from its tips A and B were determined. The crack AB extends from the tip with the minimum required critical stress which also represents the actual fracture stress of the plate. Following this procedure, it is found that the crack AB always extends from its inner tip A. The variation of the normalized critical stress σ_{2cr} of the plate versus the angle ϕ subtending between the axes of the cracks is presented in Figure 4.9 for the values of the material constant κ equal to 1.2, 1.8, 2.4 and 3.0. It is observed that σ_{2cr} decreases as the cracks rotate

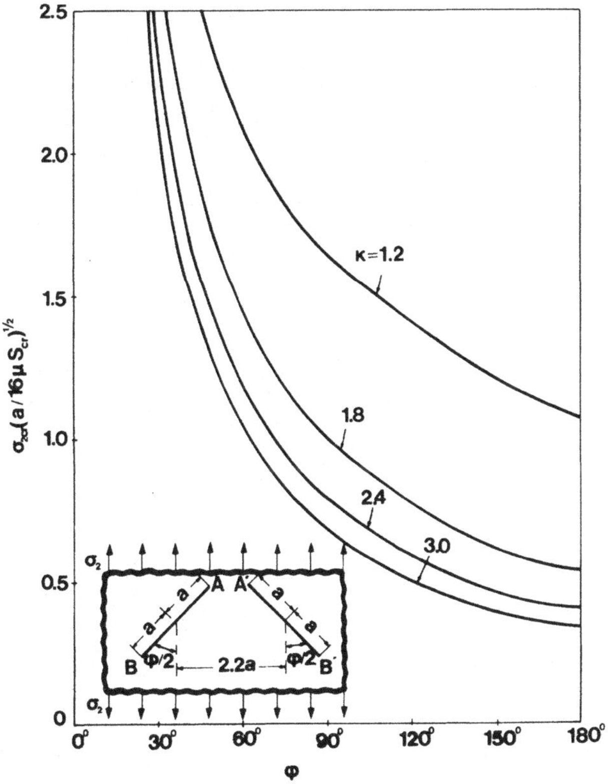

Figure 4.9. Variation of the normalized critical fracture stress σ_{2cr} of the plate of Figure 4.8 subjected to a uniform uniaxial stress σ_2 versus the angle ϕ which the crack axes subtend. The material constant κ took the values 1.2, 1.8, 2.4 and 3.0. Crack extension always starts from the inner tips A, A' of the cracks.

from a direction parallel to the applied stress σ_2 ($\phi = 0$) to a perpendicular direction ($\phi = 180°$). For $\phi = 0$ both cracks are parallel to the applied stress and the critical fracture stress σ_{2cr} becomes infinite. From Figure 4.9 we also see that σ_{2cr} decreases as κ increases: that is, when the Poisson's ratio of the material of the plate decreases. Comparing the values of σ_{2cr} with their corresponding values for the case of a single crack in an infinite plate subjected to a uniform uniaxial stress σ_2 at infinity at an angle $\phi/2$ with the crack axis, it is found that σ_{2cr} is always smaller than in the case of a single crack. Thus, the effect of the above two cracks on the strength of the plate is more dangerous than that of a single crack. This is due to the interaction effects which take place between the inner tips of the cracks, increasing the intensity of the stress field there.

The values of the angle $-\theta_{0A}$ formed by the fracture path originating from the tip A of the crack AB with the initial direction of the crack are shown in Figure 4.10. We see that θ_{0A} is always negative. This means that the new direction of the crack AB results from a clockwise rotation of its initial direction. Thus, after crack initiation, there is a tendency of the inner tips A and A' to move each toward each other. For

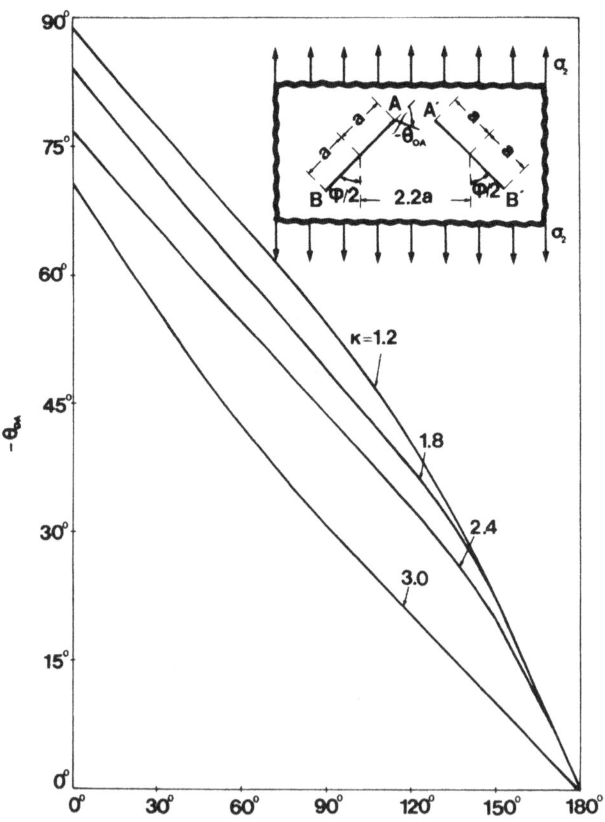

Figure 4.10. Values of the fracture angles $-\theta_{0A}$ for the case of Figure 4.9.

$\phi = 180°$ both cracks become perpendicular to the applied stress σ_2 and they propagate in their own planes ($\theta_{0A} = 0$).

b. *The stress is applied along the horizontal direction.* Let σ_1 be the applied stress and $90° - \phi/2$ the angle that it subtends with the crack axes. Working as in the previous case, the critical value of the stress σ_1 for crack extension together with the initial crack propagation pattern were determined. It was found that, contrary to the previous case, crack extension always takes place from the outer tips B and B' of the cracks. The variation of the quantity $\sigma_{1cr}(a/16\mu S_{cr})^{1/2}$ versus ϕ for $\kappa = 1.2$, 1.8, 2.4 and 3.0 is presented in Figure 4.11. We observe that as the cracks rotate from a direction perpendicular to the applied stress ($\phi = 0$) to a parallel direction ($\phi = 180°$) the stress σ_{cr} increases. This stress tends to infinity when the angle ϕ tends to 180°, that is, when the cracks become parallel to the applied stress. Comparing now the values of σ_{1cr} with the corresponding values for a single crack we observe that, contrary to the previous case of the vertically applied load, σ_{1cr} is always greater than that for a single crack. This can be attributed to the fact that in this case, where no interaction

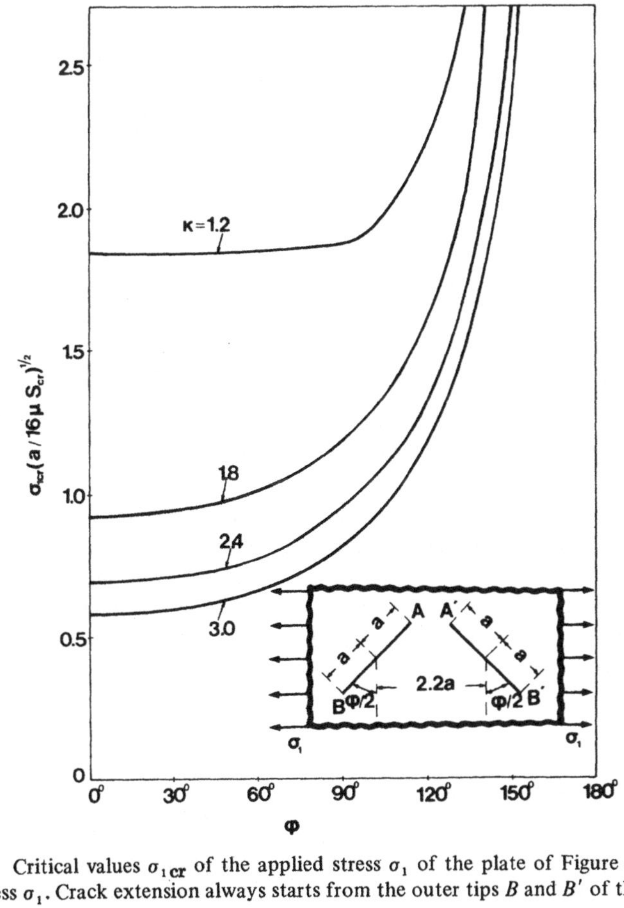

Figure 4.11. Critical values σ_{1cr} of the applied stress σ_1 of the plate of Figure 4.8 subjected to the stress σ_1. Crack extension always starts from the outer tips B and B' of the cracks.

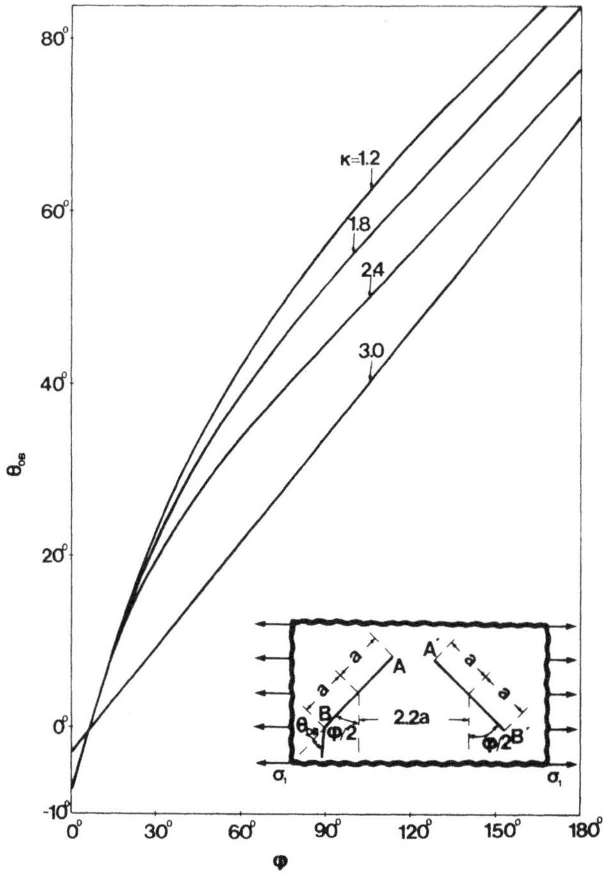

Figure 4.12. Values of the fracture angles θ_{0B} for the case of Figure 4.11.

between the inner tips takes place, the external energy imposed in the plate is equally distributed between the propagating tips B and B' and therefore larger stress is required for crack extension. This is in agreement with the well-established result that for an array of cracks perpendicularly aligned to the direction of the applied stress the critical fracture stress increases as the number of cracks increases.

The values of the angle θ_{0B} formed by the fracture path originating from the tip B of the crack AB with its initial direction are shown in Figure 4.12. We see that for $\phi > 5°$ θ_{0B} is positive, which means that the crack AB rotates after propagating in a counterclockwise manner and tends to become perpendicular to the direction of the applied stress.

4.4. Two equal inclined cracks

In this section the interaction of two equal cracks subtending an angle of inclination will be studied. Two particular configurations of the cracks will be considered. In the

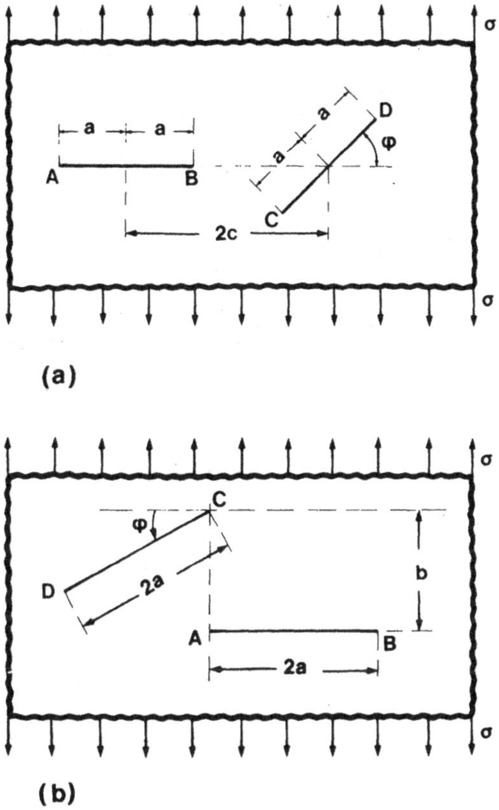

Figure 4.13. Geometrical configurations of two equal inclined cracks embedded in a plate subjected to a uniaxial stress perpendicular to one of the cracks.

first, one crack is aligned perpendicularly to the direction of the applied uniaxial stress and the center of the other passes from the extension of the first (Figure 4.13a). In the second, one crack is again perpendicular to the applied stress and one tip of the other crack passes from a point of the line perpendicular to the first crack and originating from one of its tips (Figure 4.13b). These two cases will be considered individually.

a. *The center of the second crack passes from the extension of the first* (Figure 4.13a). For this configuration of the cracks *AB* and *CD* of Figure 4.13a with a variable distance $2c$ of the centers of the cracks Hodulak and Klein [23] gave the values of k_1 and k_2 stress intensity factors for all four tips *A*, *B*, *C* and *D* of the cracks as a function of their inclination angle ϕ. From these values of k_1 and k_2, and using the strain energy density criterion, the critical values σ_{cr}^A, σ_{cr}^B, σ_{cr}^C, and σ_{cr}^D of the applied stress σ for possible brittle fracture of the plate resulting from the extension of the tips *A*, *B*, *C* and *D* were determined. It is evident that crack extension would take place from the tip of the crack *AB* or *CD* which requires the minimum critical stress for extension and that the critical stress σ_{cr} of the plate would be equal to the minimum value of the four stresses σ_{cr}^A, σ_{cr}^B, σ_{cr}^C and σ_{cr}^D. Calculating these stresses for various values of the

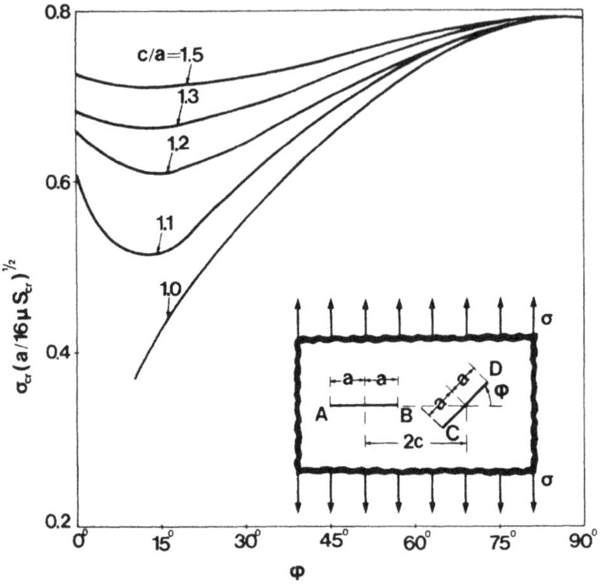

Figure 4.14. Variation of the normalized critical stress σ_{cr} of the plate of Figure 4.13a versus the angle ϕ of inclination of the cracks for $c/a = 1.0, 1.1, 1.2, 1.3$, and 1.5. The material constant κ took the value 1.8. Fracture always starts from the tip B of the crack AB.

distance c $(1.0 < c/a < 1.5)$ and the inclination angle ϕ it was found that the stress σ^B_{cr} is always the minimum critical stress. Thus, fracture of the plate would always initiate from the tip B of the crack AB that is perpendicular to the applied stress which is close to the tip C of the other crack CD. Figure 4.14 presents the variation of the quantity $\sigma_{cr}(a/16\mu S_{cr})^{1/2}$ versus the inclination angle ϕ for $c/a = 1.0$, 1.1, 1.2, 1.3 and 1.5. The material constant κ took the value 1.8. It is observed that as the crack CD moves away from the crack AB the critical stress σ_{cr} increases. Thus, the interaction effects between the two cracks are high when they are close to each other. Regarding the dependence of the stress σ_{cr} on the angle ϕ we observe that σ_{cr} becomes minimum for an angle ϕ_{cr} in the interval $(12^\circ, 15^\circ)$. Thus, for $\phi < \phi_{cr}$, σ_{cr} decreases with ϕ, while for $\phi > \phi_{cr}$ it increases. For $\phi = 90^\circ$ all curves for various values of (c/a) tend to the same value equal to 0.791, corresponding to the case of a single crack of length $2a$ in a plate loaded perpendicularly to the direction of the crack. Indeed, for $\phi = 90^\circ$ the crack CD is parallelly aligned to the direction of the applied stress and has no effect on the fracture of the plate. Comparing now the critical stress σ_{cr} of fracture of the plate with the two interacting cracks AB and CD against the corresponding stress for a plate with only the crack AB we observe that the stress σ_{cr} is always smaller for two cracks than for one.

The angles θ_{0B} formed by the fracture path originating from B with the initial direction of the crack AB are displayed in Figure 4.15. It is observed that θ_{0B} is positive except for $c/a = 1.0$ and $\phi < 30^\circ$. This means that the tip B propagates in a direction opposite to the tip C, that is, the tips B and C repel each other.

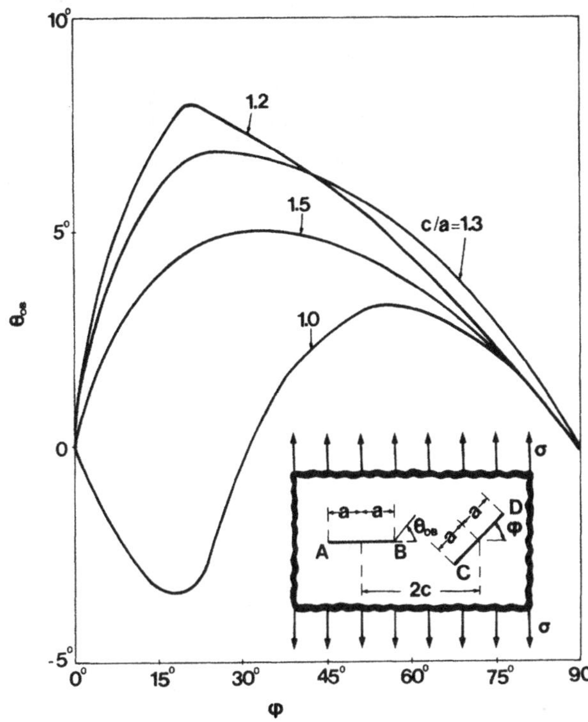

Figure 4.15. Values of the angle θ_{0B} that the fracture path originating from the tip B of crack AB in Figure 4.13a forms with the direction of AB.

b. *The one tip of the second crack passes from a point of the line perpendicular to the first crack and originating from one of its tips* (*Figure 4.13b*). This configuration of cracks has also been studied by Hodulak and Klein [23] who gave the values of k_1 and k_2 for all four tips A, B, C and D of the cracks. Working as in the previous case it was found that crack initiation starts from tip A or B of the crack AB, (which is perpendicular to the applied stress) depending on the geometry of the cracks. Figure 4.16 presents the critical values σ_{cr} of the applied stress σ versus the angle ϕ of inclination of the crack CD for $b/a = 0.4$, 0.6, 0.8, 1.0, 1.4 and 2.0. For all geometrical configurations lying above the dotted curve of the figure, fracture starts from the outer tip B of the crack AB, while for those below the dotted curve it starts from the inner tip A of AB. Thus, fracture always originates from tip B for $b/a = 2.0$ and 1.4, while for $b/a = 1.0$ it originates from B only for $\phi < -20°$. Note that negative values of the angle ϕ indicate that the crack CD lies in the opposite direction of the crack AB with respect to a line passing from C and parallel to AB. It can be observed from the above figure that for large values of the distance b, crack extension takes place from the non-interacting tip B of the crack AB, while as b decreases from the inner tip A. This is due to the fact that for small values of b the stress intensity at the tip A is highly influenced by the presence of the crack CD resulting in an increase of the magnitude of the stress field there. For large values of b, however, such interaction does not take place. Thus, the critical fracture stress σ_{cr} decreases as b decreases,

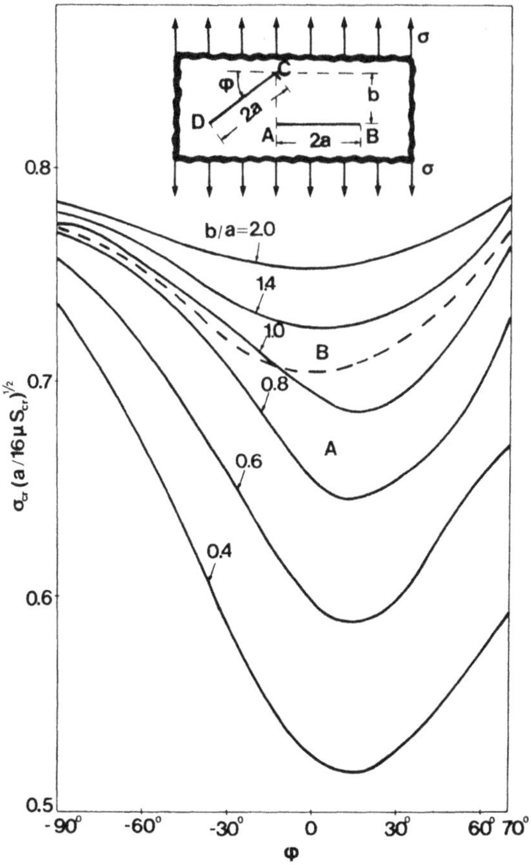

Figure 4.16. Variation of the normalized critical stress σ_{cr} of the plate of Figure 4.13b versus the angle ϕ of inclination of the cracks for $b/a = 0.4$, 0.6, 0.8, 1.0, 1.4 and 2.0. Areas where fracture of the plate takes place from the tip A or B of the crack AB are separated by the dotted curve of figure.

or in other words, as the two cracks approach each other. Regarding the dependence of the stress σ_{cr} on the angle ϕ it is observed that σ_{cr} takes its minimum value in the interval (0°, 30°) and depends on the value of b. Thus, for $\phi < \phi_{cr}$, where ϕ_{cr} corresponds to $(\sigma_{cr})_{min}$, σ_{cr} decreases with ϕ, while for $\phi > \phi_{cr}$ it increases. Thus, the most dangerous configuration of the two cracks is that in which CD slightly deviates from the parallel to AB and leans to the direction of AB. The values of the angle $\theta_{0A,B}$ of initial crack extension for the case of Figure 4.16 are shown in Figure 4.17.

4.5. X-formed arrays of cracks

In this section the interaction of cracks arranged along two radial lines in an infinite plate will be studied. Two particular configurations of cracks shown in Figure 4.18

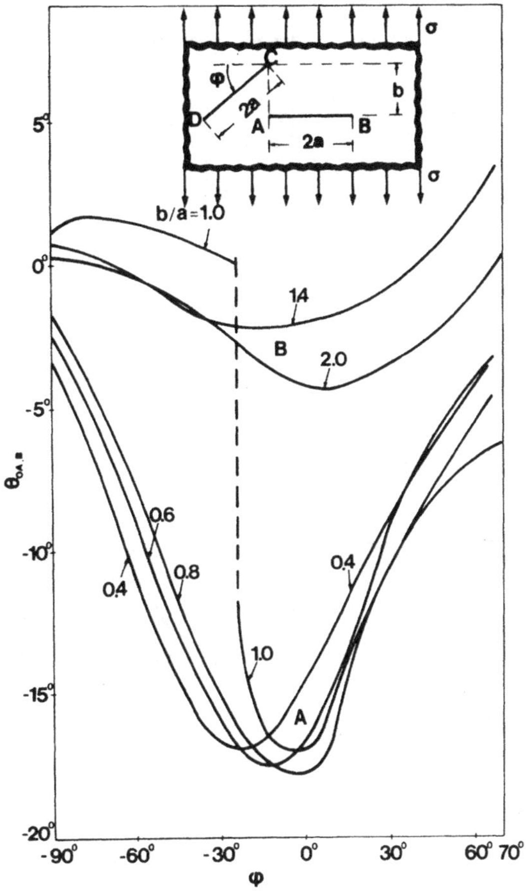

Figure 4.17. Values of the fracture angles $\theta_{0B,C}$ for Figure 4.13b.

are considered. In the first, (Figure 4.18a), the plate contains three equal cracks, two of which lie on an inclined line passing through the center of the third. In the second, (Figure 4.18b), two unequal inclined cracks cross each other. For both these cases the brittle fracture behavior of the plate, incorporating the determination of the critical load for crack extension and the initial pattern of crack propagation will be studied.

a. *Three equal cracks lying on two inclined crossing lines* (*Figure 4.18a*): This crack arrangement is embedded in an infinite isotropic elastic plate loaded by a uniform all-round tension σ. For this case and for a variable angle of inclination ϕ of the axes of the cracks Aksogan [22] gave the values of the k_1 and k_2 stress intensity factors for the crack tips X, Y and Z. Using the strain energy density criterion the critical stresses σ_{cr}^X, σ_{cr}^Y and σ_{cr}^Z for possible initiation of fracture from the tips X, Y and Z were determined. It was found that for $\phi < 32°$ the stress σ_{cr}^X is smaller than the stresses σ_{cr}^Y and σ_{cr}^Z. Therefore, fracture starts from the tip X and the stress σ_{cr}^X represents the critical fracture stress of the plate. For the remaining values of ϕ ($\phi > 32°$)

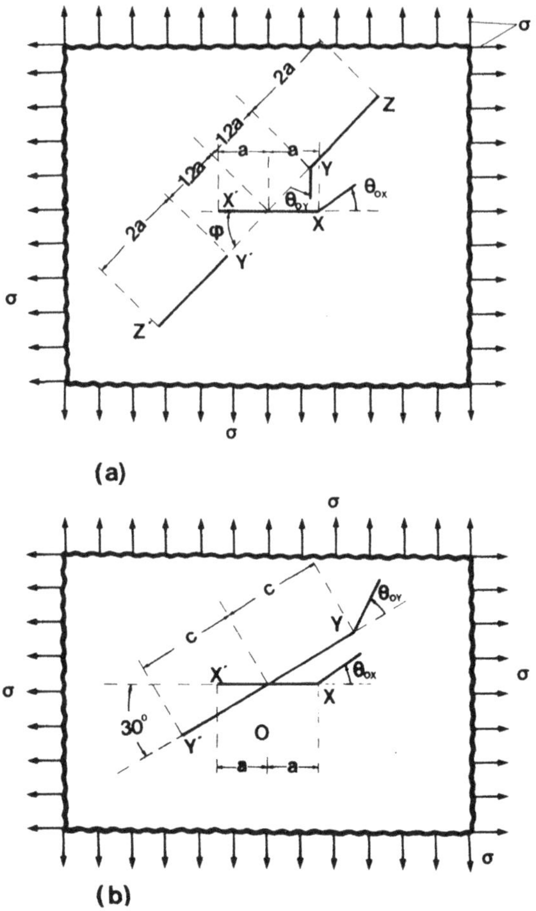

Figure 4.18. Geometry of two configurations of X-formed arrays of cracks.

fracture starts from the tip Y of the other crack YZ. It is thus observed that fracture of the plate always starts from the interacting inner tips X and Y of the cracks. The variation of the normalized critical stress σ_{cr} versus the angle ϕ of inclination of the axes of the cracks is shown in Figure 4.19 for $\kappa = 1.2$, 1.8, 2.4 and 3.0. In this figure a dotted vertical line passing from the point $\phi = 32^\circ$ separates the areas in which fracture starts from the tip X or Y. It is observed that σ_{cr} becomes minimum for $\phi = 10^\circ$, and in the interval (0, 10°) σ_{cr} decreases with ϕ, while for $\phi > 10^\circ$, σ_{cr} increases as ϕ increases. The values of the initial fracture angles θ_{0X} and θ_{0Y} are presented in Figure 4.20.

b. *Two unequal inclined cracks* (*Figure 4.18b*): For this case and for an all-round uniform tension σ the values of k_1 and k_2 for the tips X and Y were given by Aksogan [22]. Using the strain energy density criterion it was found that for $c/a < 0.88$ crack extension starts from the tip X of the crack XX', while for the remaining values of c/a it starts from the tip Y. The variation of the normalized critical stress σ_{cr} versus

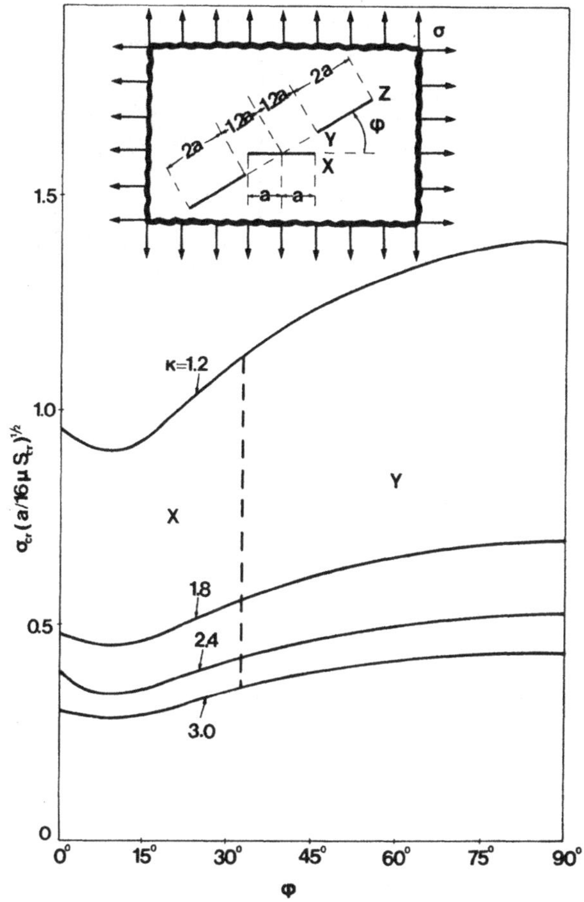

Figure 4.19. Variation of the normalized critical fracture stress σ_{cr} for the case of Figure 4.18a versus the angle ϕ of the inclination of the cracks for $\kappa = 1.2$, 1.8, 2.4 and 3.0. The dotted line separates areas in which fracture starts from the tips X and Y.

c/a is presented in Figure 4.21 for $\kappa = 1.2$, 1.8, 2.4 and 3.0. In this figure a dotted line was drawn to separate areas in which fracture starts from the tips X and Y. It is observed that the $(c/a) - \sigma_{cr}$ curves are bell-shaped with the stress σ_{cr} becoming maximum when the lengths of the cracks are equal. For $c = 0$ the values of critical stress coincide with those of a single straight crack. They are 1.581, 0.791, 0.598 and 0.5 for $\kappa = 1.2$, 1.8, 2.4 and 3.0 respectively. As the length c of the inclined crack YY' increases up to the value $c = 0.88a$ the critical stress required for fracture of the plate also increases. This can be explained by the observation that when the lengths of the cracks become comparable the imposed external energy of the plate is shared between the tips of the cracks, and therefore larger external stress is required for crack propagation. As c becomes greater than a the most part of this energy is absorbed by the tip Y of the larger crack and so the critical stress

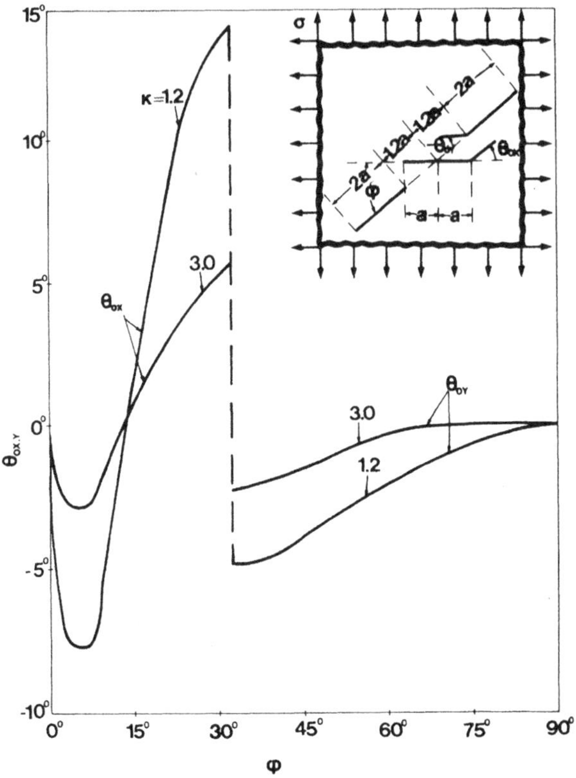

Figure 4.20. Values of the fracture angles $\theta_{0X,Y}$ for the case of Figure 4.19.

decreases. The values of the crack extension angles θ_{0X} and θ_{0Y} are presented in Figure 4.22.

References

[1] Westergaard, H.M., Bearing pressures and cracks, *Journal of Applied Mechanics*, 6, *Trans. ASME Ser. E*, pp. A49–53 (1939).

[2] Willmore, T.J., The distribution of stress in the neighborhood of a crack, *Quarterly Journal of Mechanics and Applied Mathematics*, 2, pp. 53–63 (1949).

[3] Irwin, G.R., Analysis of stresses and strains near the ends of a crack traversing a plate, *Journal of Applied Mechanics*, Trans. ASME, 24, pp. 361–364 (1957).

[4] Irwin, G.R., Fracture, *Handbuch der Physik*, Volume VI, Springer Verlag, pp. 551–590 (1958).

[5] Panasyuk, V.V. and Lozovoi, B.L., On the development of two cracks of unequal length, *Dopovidi Akademiia Nauk Ukranian RSR*, 11, pp. 1444–1447 (1962).

[6] Erdogan, F., On the stress distribution in plates with collinear cuts under arbitrary loads, *Proceedings of the Fourth U.S. National Congress of Applied Mechanics*, pp. 547–553 (1962).

[7] Smith, E., The spread of plasticity between two cracks, *International Journal of Engineering Science*, 2, pp. 379–387 (1964).

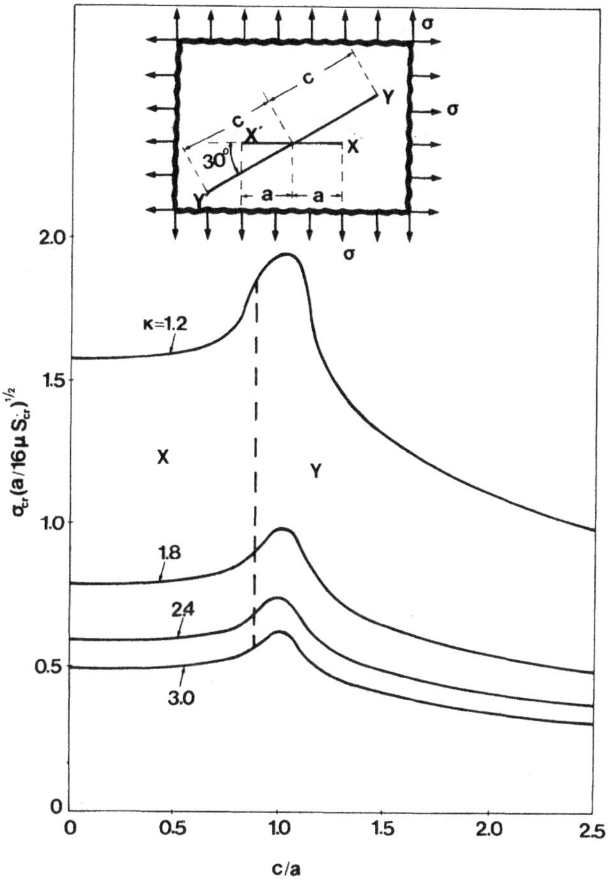

Figure 4.21. Variation of the normalized critical fracture stress σ_{cr} for the case of Figure 4.18b versus the ratio c/a of the lengths of the cracks. Areas in which fracture starts from the tips X and Y are separated by dotted line.

[8] Smith, E., The spread of plasticity from stress concentration, *Proceedings of the Royal Society*, Series A, 282, pp. 422–432 (1964).

[9] Smith, E., Fracture at stress concentration, *Proceedings of the First International Conference of Fracture*, 1, pp. 133–151 (1966).

[10] Bereznicki, L.T., Stress concentration, *Academy of Ukranian SSR Institute of Mechanics*, 1 (1965).

[11] Yokobori, T.Y., Ichikawa, M. and Ohashi, M., Interaction between elastic cracks, dislocation cracks and slip bands, *Proceedings of the First International Conference of Fracture*, 1, pp. 167–190 (1966).

[12] Theocaris, P.S., Interaction of cracks with other cracks or boundaries *Materialprüfung*, 13, pp. 264–269 (1971).

[13] Theocaris, P.S., Interaction between collinear asymmetric cracks, *Journal of Strain Analysis*, 7, pp. 186–193 (1972).

[14] Theocaris, P.S., Constrained zones in a periodic array of collinear equal cracks, *International Journal of Mechanical Sciences*, 14, pp. 79–94 (1972).

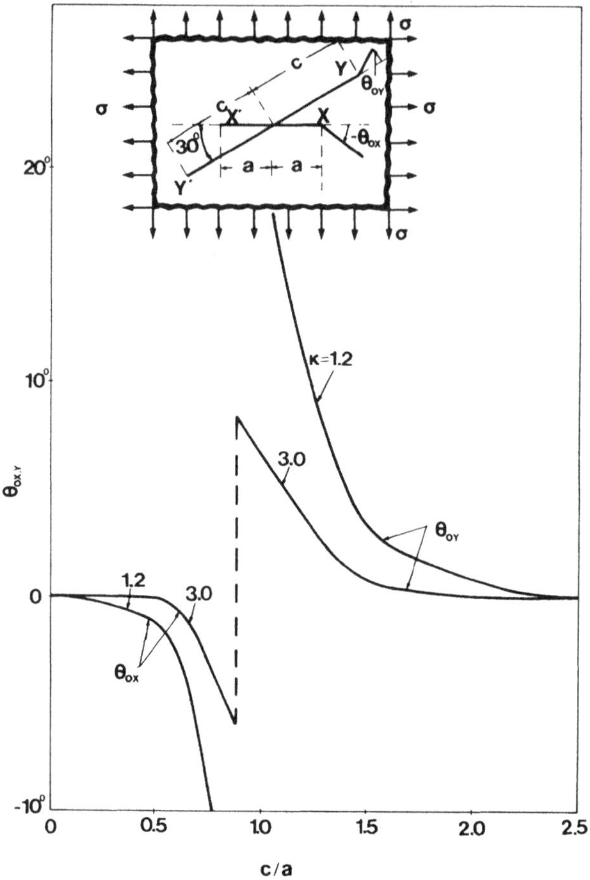

Figure 4.22. Values of the fracture angles $\theta_{0X,Y}$ for the case of Figure 4.21.

[15] Theocaris, P.S., A theoretical consideration of the constrained zones in an array of interacting collinear and asymmetric cracks, *Acta Mechanica*, 17, pp. 169–189 (1973).

[16] Lange, F.F., Interaction between overlapping parallel cracks; a photoelastic study, *International Journal of Fracture Mechanics*, 4, pp. 287–294 (1968).

[17] Yokobori, T., Ouzumi, M. and Ichikawa, M., Strength and Fracture of Materials, *Report of the Research Institute*, Tohoku University (1971).

[18] Isida, M., Analysis of stress intensity factors for plates containing random array of cracks, *Bulletin of the Japanese Society of Mechanical Engineers*, 13, pp. 635–642 (1970).

[19] Isida, M., Method of Laurent series expansion for internal crack problems, *Methods of Analysis and Solutions of Crack Problems*, Edited by G.C. Sih, Noordhoff International Publishing, Leyden, pp. 56–130 (1973).

[20] Datsyshin, A.P. and Suvruk, M.P., A system of arbitrarily oriented cracks in elastic solids, *Prikladnaya Matematika y Mekhanika (Journal of Applied Mathematics and Mechanics)*, 37, pp. 326–332 (1973).

[21] Aksogan, O., The interaction of collinear arrays of Griffith cracks on two radial lines, *Journal of Engineering for Industry, Trans. ASME*, pp. 1086–1091 (1976).

[22] Aksogan, O., The stress intensity factors for X-formed arrays of cracks, *Advances in Research on the Strength and Fracture of Materials*, Editor D.M.R. Taplin, Vol. 3A, *Analysis and Mechanics*, Pergamon Press, pp. 177–184 (1977).

[23] Hodulak, L. and Klein, G., Spannungsanalyse von Mehrfachrissen (Interaction of Multiple Cracks), *Institut für Festkörpermechanik Report*, Freiburg, December (1974).

[24] Badaliance, R. and Gupta, G.D., Growth characteristics of two interacting cracks, *Engineering Fracture Mechanics*, 8, pp. 341–353 (1976).

[25] Swain, M.V. and Hagan, J.T., Some observations of overlapping interacting cracks, *Engineering Fracture Mechanics*, 10, pp. 299–304 (1978).

[26] Sih, G.C., *Handbook of Stress Intensity Factors*, Institute of Fracture and Solid Mechanics, Lehigh University, pp. 1.2.6-1–1.2.6-9 (1973).

[27] Rooke, D.P. and Cartwright, D.J., *Stress Intensity Factors*, Hillingdon Press, London, pp. 120–124 (1976).

[28] Theocaris, P.S., Dynamic propagation and arrest measurements by the method of caustics on overlapping skew-parallel cracks, *International Journal of Solids and Structures*, 14, pp. 639–653 (1978).

[29] Gdoutos, E.E., Interaction between two equal skew-parallel cracks, *Journal of Strain Analysis*, 15, pp. 127–136 (1980).

5

Arc-shaped cracks

5.1. Introduction

This chapter is devoted to the study of the brittle fracture behavior of plates weakened by arc-shaped cracks. Cracks with curved fronts are frequently encountered in practical applications, generally resulting from the propagation of straight cracks subjected to mixed-mode loading. The plane elastostatic problem of a circular crack in an infinite plate was first studied by Muskhelishvili [1] who used his complex variable approach. He gave the expressions of the complex potentials for the situation when the chord of the circular crack is arbitrarily oriented with respect to the direction of the applied uniaxial or biaxial stress at infinity. Sih *et al.* [2] solved this problem and gave the expressions of opening-mode and sliding-mode stress intensity factors at the tips of the circular crack.

In the present chapter the strain energy density criterion is used to analyze the brittle fracture characteristic quantities of a plate with circular cracks. The cases of a single and a periodic array or a star-shaped array of circular cracks are considered, and the critical loads for failure initiation as well as the initial pattern of crack extension are determined.

5.2. The circular crack under uniform stress

Consider a circular crack AA' of radius R and angle 2β in an infinite isotropic elastic plate loaded by a uniform uniaxial stress σ at infinity (Figure 5.1). The cases when the chord of the crack is perpendicular or parallel to the applied stress will be considered separately.

a. *The chord of the crack is perpendicular to the applied stress* (Figure 5.1a). Consider a Cartesian coordinate system attached to the crack tip A with its x-axis coinciding with the tangent of the crack at A. The stress field in the vicinity of A is given by the general equations (2.1) of a crack in a mixed-mode stress field. The values of the k_1 and k_2 stress intensity factors are given by [2]:

$$k_1 = \frac{\sigma}{2}(R\sin\beta)^{1/2}\left[\frac{\left(1-\sin^2\frac{\beta}{2}\cos^2\frac{\beta}{2}\right)\cos\frac{\beta}{2}}{1+\sin^2\frac{\beta}{2}} + \cos\frac{3\beta}{2}\right] \tag{5.1a}$$

$$k_2 = \frac{\sigma}{2}(R\sin\beta)^{1/2}\left[\frac{\left(1-\sin^2\frac{\beta}{2}\cos^2\frac{\beta}{2}\right)\sin\frac{\beta}{2}}{1+\sin^2\frac{\beta}{2}} + \sin\frac{3\beta}{2}\right]. \tag{5.1b}$$

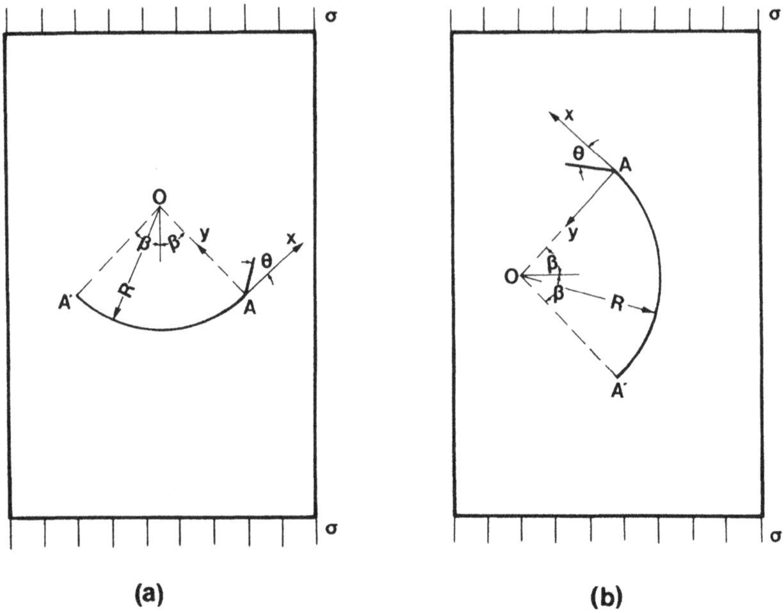

Figure 5.1. Geometry of a plate containing a circular crack with its chord perpendicular (a) or parallel (b) to the direction of the applied uniaxial stress.

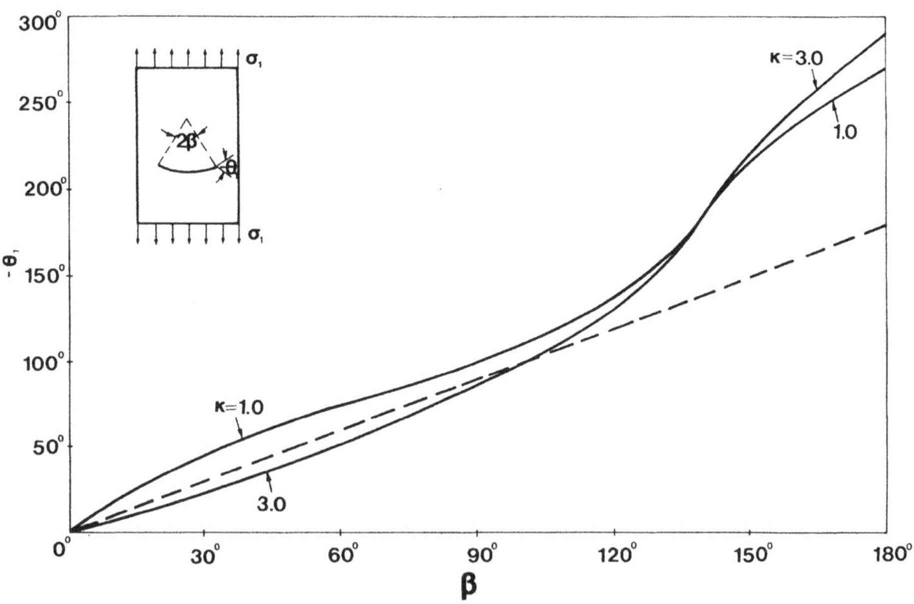

Figure 5.2. Variation of the crack extension angle $-\theta_1$ with the half angle β of the circular crack whose chord is perpendicular to the applied tensile stress for $\kappa = 1.0$ and 3.0. The straight line corresponding to the extension of the crack at right angles to the direction of the applied load is also indicated (dotted curve).

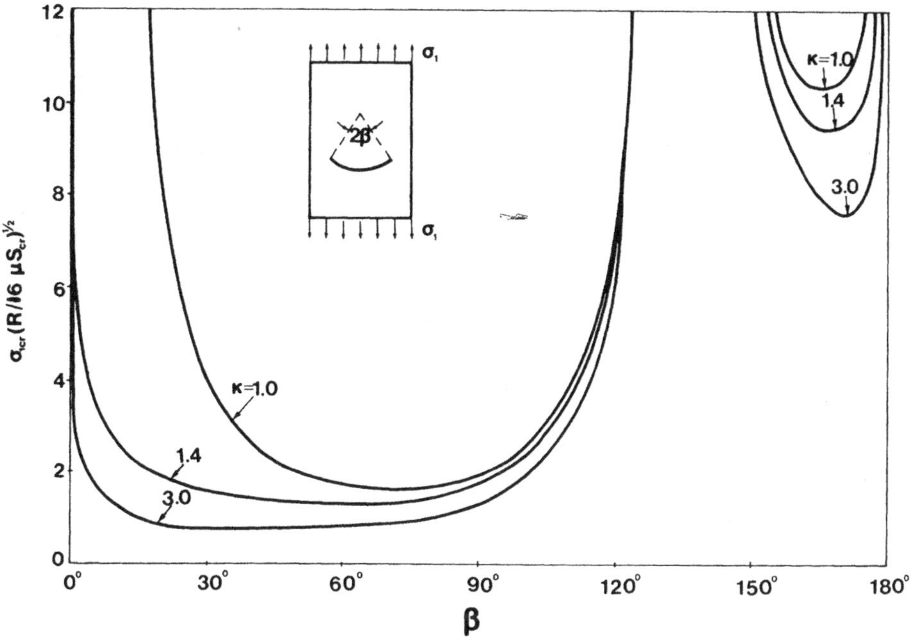

Figure 5.3. Variation of the normalized critical stress σ_{1cr} for crack extension with the half angle β of the circular crack whose chord is perpendicular to the applied tensile stress for $\kappa = 1.0$, 1.4 and 3.0.

When the values of k_1 and k_2 are introduced into equation (2.6a) of the strain energy density criterion an equation containing the quantities β, κ and θ is obtained. Solution of this equation in conjunction with inequality (2.6b) gives the values of the crack extension angle θ_0. If now θ_0 is introduced into equation (2.4) the minimum value S_{min} of the strain energy density factor S is obtained. S_{min} is then equated to the critical value S_{cr} of S in order to calculate the critical value σ_{cr} of the applied stress σ for crack extension.

Figure 5.2 presents the variation of the crack extension angle $-\theta_1$ versus the half angle of the circular crack β when the applied load is tensile and for the extreme values of κ equal to 1.0 and 3.0. The curves corresponding to the intermediate values of κ have been omitted to preserve clarity of the figure. In the same figure the straight line $-\theta_1 = \beta$ corresponding to extension of the crack at a right angle to the direction of the applied load has also been plotted. We see that the angle $-\theta_1$ increases monotonically with β, and that in the interval $0° < \beta < 137.5°$ the angles $-\theta_1$ for $\kappa = 1.0$ are always greater than for $\kappa = 3.0$. This rule is reversed in the interval $137.5° < \beta < 180°$. Furthermore, Figure 5.2 illustrates that initial crack extension takes place in a direction almost normal to the applied load for all values of β in the interval $0° < \beta < 120°$. When the angle β is greater than $120°$ the direction of crack extension deviates from that normal to the load, becoming parallel to the applied load for $\beta = 180°$. The values of the quantity $\sigma_{1cr}(R/16\mu S_{cr})^{1/2}$ for $\kappa = 1.0$, 1.4 and 3.0 are presented in Figure 5.3. It can be seen that all curves for the values of the angle β equal to $\beta = 0°$, $180°$ and

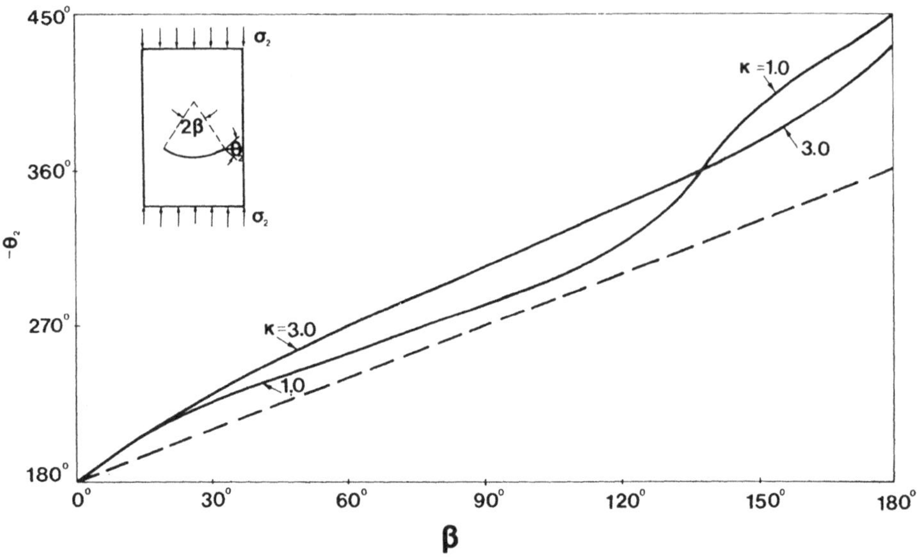

Figure 5.4. As in Figure 5.2 for compressive applied stress.

137.5° tend to infinity. It is thus concluded that a plate weakened by a circular crack requires an infinitely large stress for crack extension not only for the trivial case of zero angle circular crack but also for the values of $2\beta = 275°$ and 360°. The infinite value of stress for crack extension is given by the concept of linear theory of fracture. Its physical meaning is that failure of the cracked plate takes place at the same critical load as failure of the uncracked plate. We further observe that the critical stress for crack extension decreases as κ increases, or as the Poisson's ratio ν decreases. We also see that for each value of κ there is a specific value of the angle of crack 2β at which the required stress for crack extension reaches minimum. This critical value of the angle 2β is equal to 85°, 115° and 140° for $\kappa = 3.0$, 1.4 and 1.0 respectively.

Figures 5.4 and 5.5 present the variation of the quantities $-\theta_2$ and $\sigma_{2\,\mathrm{cr}}(R/16\mu S_{\mathrm{cr}})^{1/2}$ versus β when the applied load is compressive. In Figure 5.4 the straight line $-\theta_2 = (180° + \beta)$ corresponding to the crack extension normal to the direction of the applied load has also been plotted. As we observe from this figure the direction of initial crack extension approaches the direction normal to the applied load for the values of β in the interval $0° < \beta < 120°$, while it is almost parallel to the load for values of β greater than 120°. Contrary to application of tensile loads, where there is a value of the angle β in the interval $0° < \beta < 180°$ for which an infinitely large stress is required for crack extension, no such value exists for compressive loads. Further considerations analogous to those made previously for tensile loads can be established from Figures 5.4 and 5.5.

The following conclusions may be drawn concerning the dependence of the angle of initial crack extension and the critical fracture stress on the length of the circular crack and type of the applied load:

(i) For tensile applied loads, the direction of initial crack extension is always

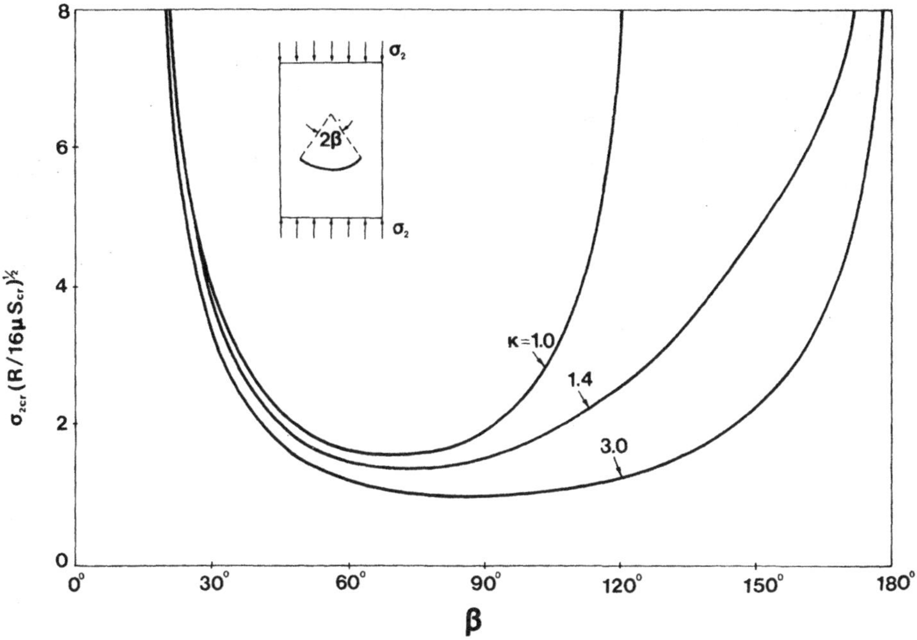

Figure 5.5. As in Figure 5.3 for compressive applied stress.

outside the angle formed by the positive tangent (indicating the direction of increasing crack angles 2β) and the line connecting the center with the tip of the crack. For compressive loads however, crack extension takes place in the opposite direction. The crack propagation direction for both tensile and compressive loads is approximately perpendicular to the direction of the load for crack angles 2β less than 240°, while for angles 2β greater than 240° the crack extends close to the direction of the load. The angle of crack extension $-\theta_1$ for tensile applied loads with crack angles 2β less than 275° increases with Poisson's ratio ν, while it decreases with ν for angles 2β greater than 275°. Quite the contrary occurs for compressive loads. Furthermore, for both tensile and compressive loads, angles $-\theta_{1,2}$ increase with the crack angle 2β.

(ii) The presence of a circular crack of angle 360° in a plate does not reduce its fracture stress for either tensile or compressive loads. Furthermore, under tensile loads the critical stress for crack propagation becomes infinitely large for the value of the crack angle 2β equal to 275°. Such a crack angle does not exist within the interval $0° < 2\beta < 360°$ for compressive loads. The critical stress $\sigma_{1,2cr}$ for crack extension under both tensile and compressive loads decreases as the Poisson's ratio of the plate material decreases. For compressive loads there exists for each value of ν a specific value of the crack angle 2β at which the critical stress σ_{2cr} becomes minimum, while for tensile loads there are two such local minima in the intervals $0° < 2\beta < 275°$ and $275° < 2\beta < 360°$ (See Figures 5.3 and 5.5).

b. *The chord of the crack is parallel to the applied stress (Figure 5.1b).* The values of k_1 and k_2 stress intensity factors are given by [2]:

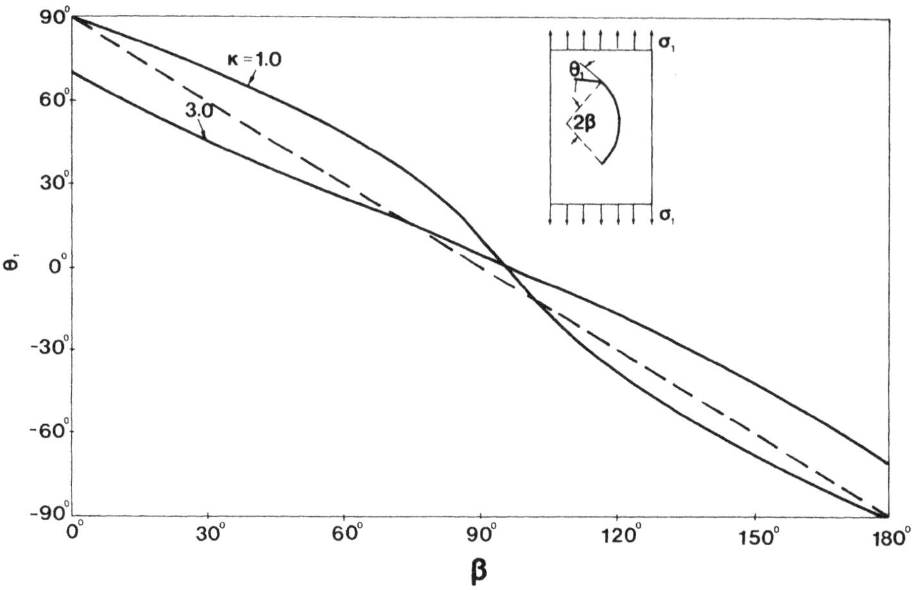

Figure 5.6. Variation of the crack extension angle θ_1 with the half angle β of the circular crack whose chord is parallel to the applied tensile stress for $\kappa = 1.0$ and 3.0. The straight line corresponding to the extension of the crack at right angles to the direction of the applied load is also indicated (dotted curve).

$$k_1 = \frac{\sigma}{2}(R \sin\beta)^{1/2}\left[\frac{\left(1 + \sin^2\frac{\beta}{2}\cos^2\frac{\beta}{2}\right)\cos\frac{\beta}{2}}{1 + \sin^2\frac{\beta}{2}} - \cos\frac{3\beta}{2}\right] \tag{5.2a}$$

$$k_2 = \frac{\sigma}{2}(R \sin\beta)^{1/2}\left[\frac{\left(1 + \sin^2\frac{\beta}{2}\cos^2\frac{\beta}{2}\right)\sin\frac{\beta}{2}}{1 + \sin^2\frac{\beta}{2}} - \sin\frac{3\beta}{2}\right]. \tag{5.2b}$$

Proceeding as in the previous case, the crack extension angle $\theta_{1,2}$ and the critical stress $\sigma_{1,2\text{cr}}$ for crack propagation were determined. The variation of the quantities $\theta_{1,2}$ and $\sigma_{1,2\text{cr}}$ versus β is shown for tensile loads in Figures 5.6 and 5.7 and for compressive loads in Figures 5.8 and 5.9. These figures illustrate the following conclusions similar to those noted for the circular crack with its chord perpendicular to the direction of the applied load regarding the direction of initial crack extension and the critical applied stress:

(i) The initial direction of crack extension is approximately perpendicular to the direction of the applied tensile or compressive load for all values of the crack angle 2β. For values of the angle 2β less than 190° the crack extension angles θ_1 and $-\theta_2$ for tensile and compressive loads decrease as the Poisson's ratio ν decreases, while for 2β greater than 190° this rule is reversed. Furthermore, angles $-\theta_1$ and $-\theta_2$ increase with β.

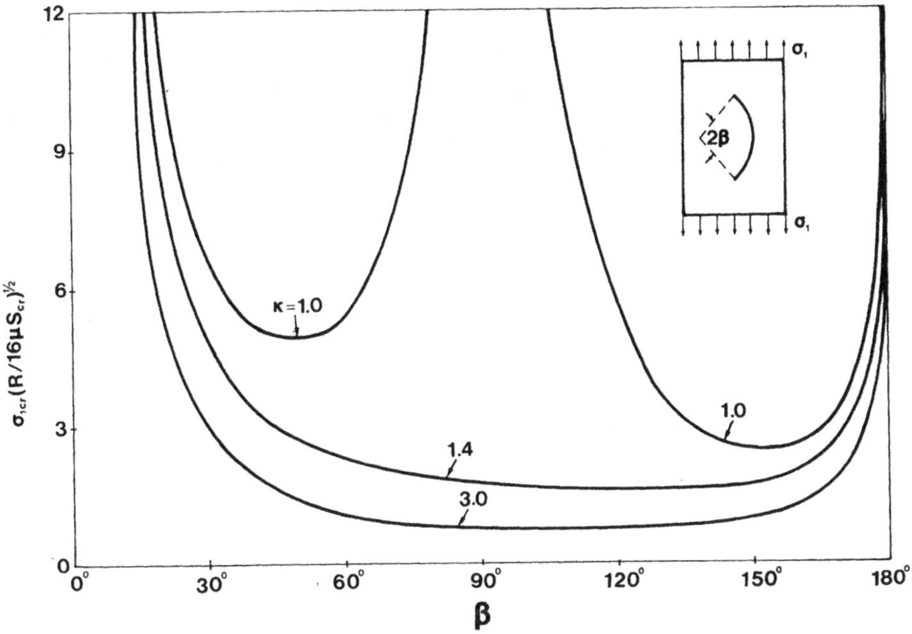

Figure 5.7. Variation of the normalized critical stress σ_{1cr} for crack extension with the half angle β of the circular crack whose chord is parallel to the applied tensile stress for $\kappa = 1.0$, 1.4 and 3.0.

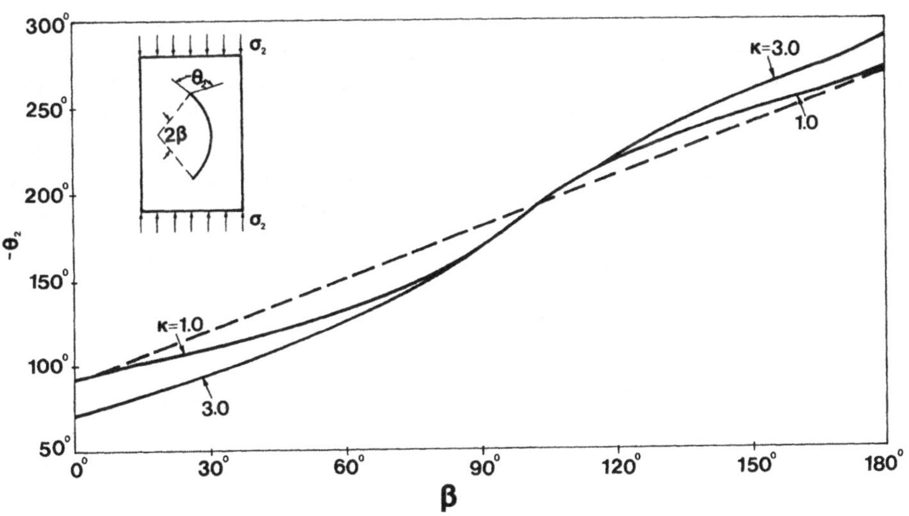

Figure 5.8. As in Figure 5.6 for the angle $-\theta_2$ and compressive applied stress.

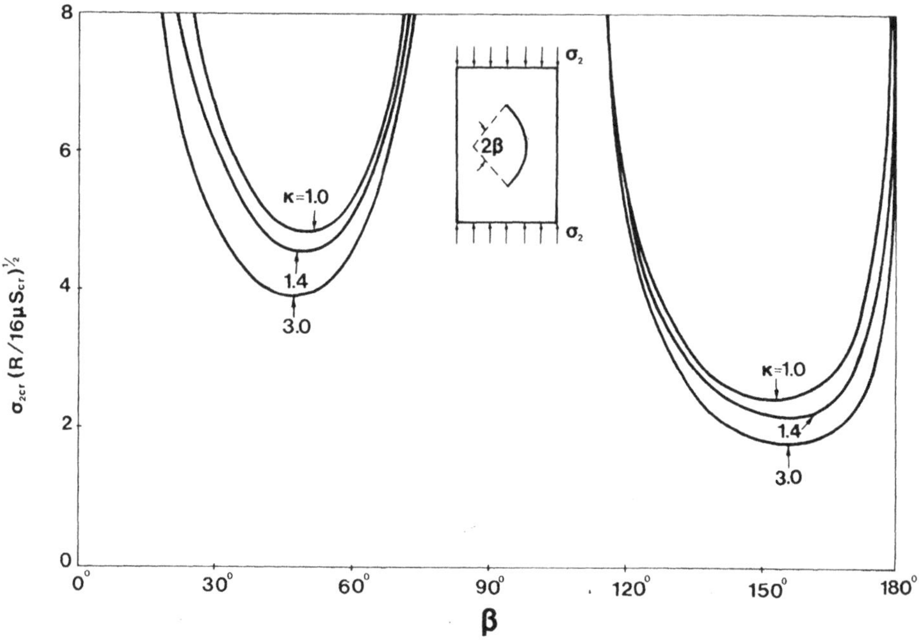

Figure 5.9. As in Figure 5.7 for compressive applied stress.

(ii) As in the previous case, the presence of a circular crack of angle 360° in a plate does not reduce its strength for either tensile or compressive loads. Furthermore, under compressive loads the strength of the cracked plate is unaffected by a circular crack of angle 190° for all values of Poisson's ratio of the plate. However, for tensile loads, a circular crack of angle 190° lowers the strength of the plate except when the plate is under plane-strain conditions and its Poisson's ratio ν is equal to 0.5.

The critical stress $\sigma_{1,2cr}$ for crack propagation decreases as the Poisson's ratio ν of the plate material decreases under both tensile and compressive loads. For tensile loads at each value of ν there exists a specific value of the crack angle 2β for which the critical stress σ_{1cr} reaches minimum (except $\kappa = 1.0$, for which there are two such minima). For compressive loads there are two such minimum values of σ_{2cr} in the intervals $0° < 2\beta < 190°$ and $190° < 2\beta < 360°$. The derivations of this section were taken from reference [3].

5.3. A periodic array of circular cracks

In this section the brittle fracture behavior of a periodic array of circular cracks in an elastic plate will be studied. The geometry of the cracks is shown in Figures 5.10a and 5.10b. In Figure 5.10a the line passing through the centers of the cracks is perpendicular, while in Figure 5.10b it is parallel to the direction of the applied tensile stress. These two problems have been considered by Ioakimidis and Theocaris [4], who gave the values of the k_1 and k_2 stress intensity factors at the crack tips. The critical stress

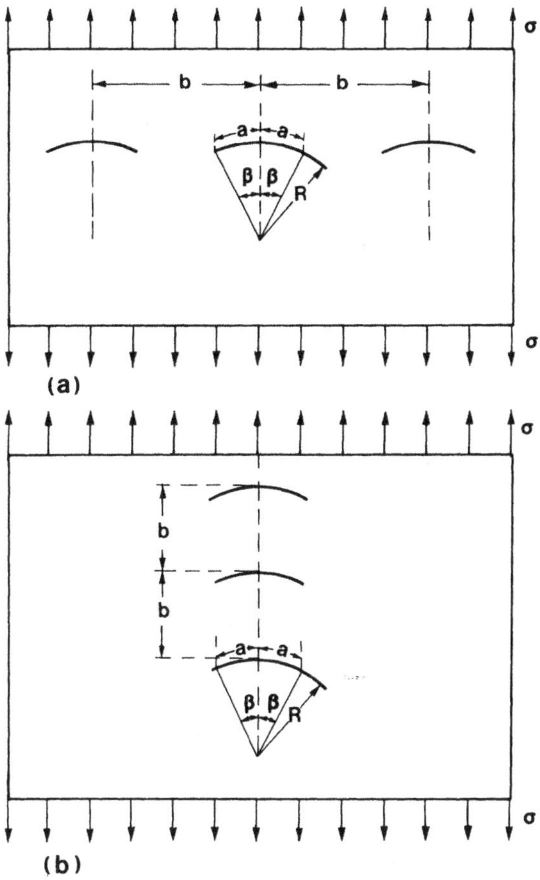

Figure 5.10. Geometrical configuration of a periodic array of circular cracks with the line passing through their centers perpendicular (a) or parallel (b) to the direction of the applied stress.

for crack extension and the initial pattern of crack propagation are determined by using the values of k_1 and k_2 in combination with the strain energy density criterion.

Figure 5.11 presents the variation of the quantity $\sigma_{cr}(a/16\mu S_{cr})^{1/2}$ versus the ratio $2a/b$ of the crack length $2a$ to the period of the array b for values of the half crack angles β equal to 0, 15°, 30°, 45°, 60° and 75°. The material constant κ is equal to 1.8. The limiting value of $(2a/b) = 0$ corresponds to a single circular crack loaded normally to its chord. The values of the quantity $\sigma_{cr}(a/16\mu S_{cr})^{1/2}$ for $2a/b = 0$ are equal to those found in Section 5.2a. Figure 5.11 reveals that the critical stress for the array of cracks is always smaller than that for a single circular crack. This can be attributed to the interaction effects between the cracks which result in a reduction of the applied critical stress. Furthermore, we observe from Figure 5.11 that the stress σ_{cr} decreases as the crack angle β decreases. Thus, for each value of $2a/b$, σ_{cr} reaches minimum for $\beta = 0°$, corresponding to the case of an array of straight cracks perpendicular to the

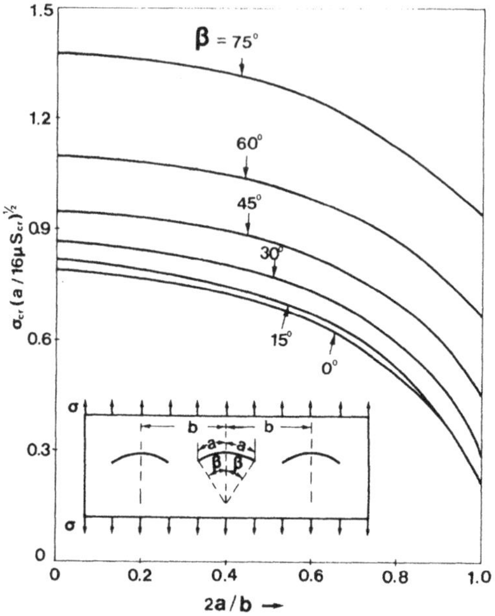

Figure 5.11. Variation of the quantity $\sigma_{cr}(a/16\mu S_{cr})^{1/2}$ versus the ratio $2a/b$ of the crack length $2a$ to the period b of the array of cracks of Figure 5.10a for $\beta = 0, 15°, 30°, 45°, 60°$ and $75°$. The material constant κ takes the value 1.8.

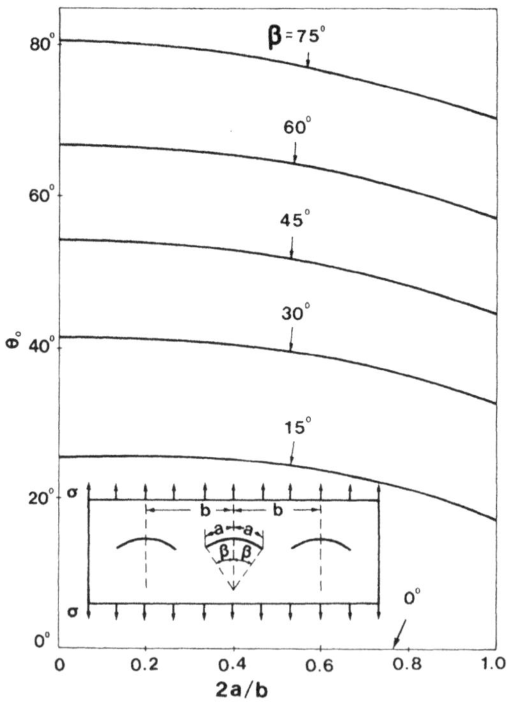

Figure 5.12. Values of the crack extension angle θ_0 corresponding to the right tip of the array of cracks of Figure 5.10a.

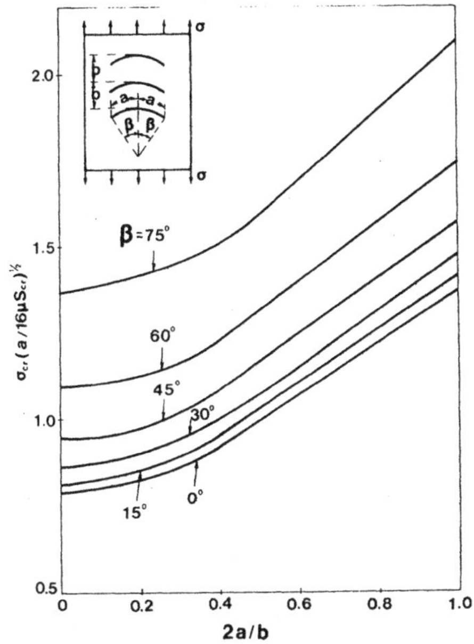

Figure 5.13. Variation of the quantity $\sigma_{cr}(a/16\mu S_{cr})^{1/2}$ versus the ratio $2a/b$ of the crack length $2a$ to the period b of the array of cracks of Figure 5.10b for $\beta = 0$, 15°, 30°, 45°, 60° and 75°. The material constant κ takes the value 1.8.

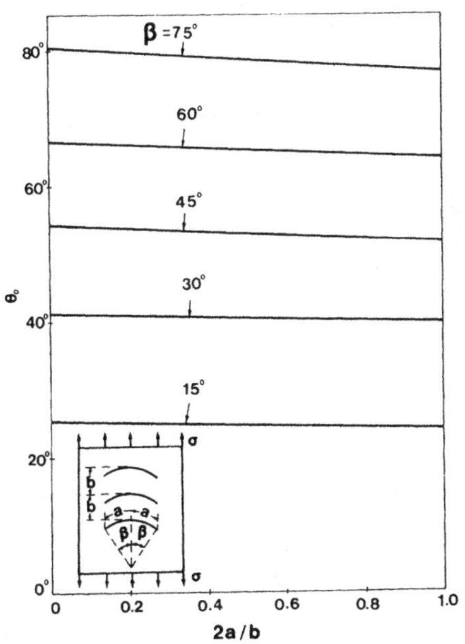

Figure 5.14. Values of the crack extension angle θ_0 corresponding to the right tip of the array of cracks of Figure 5.10b.

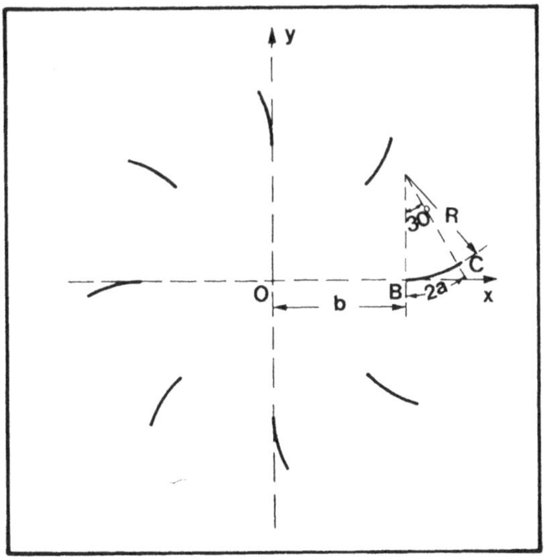

Figure 5.15. Geometry of a periodic star-shaped array of circular cracks.

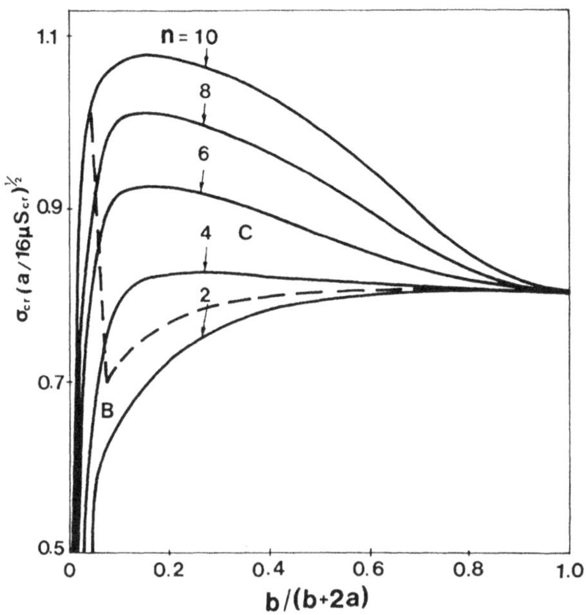

Figure 5.16. Variation of the normalized critical stress $\sigma_{cr}(a/16\mu S_{cr})^{1/2}$ versus the non-dimensional quantity $b/(b + 2a)$ for the case of Figure 5.15. Areas in which crack extension starts from the inner, B, or the outer, C, tips of the cracks are separated by a dotted line. The material constant κ takes the value 1.8.

applied stress. The existence, therefore, of a curvature in the cracks always increases the critical stress.

The variation of the angle θ_0 formed by the fracture path starting from the right tip

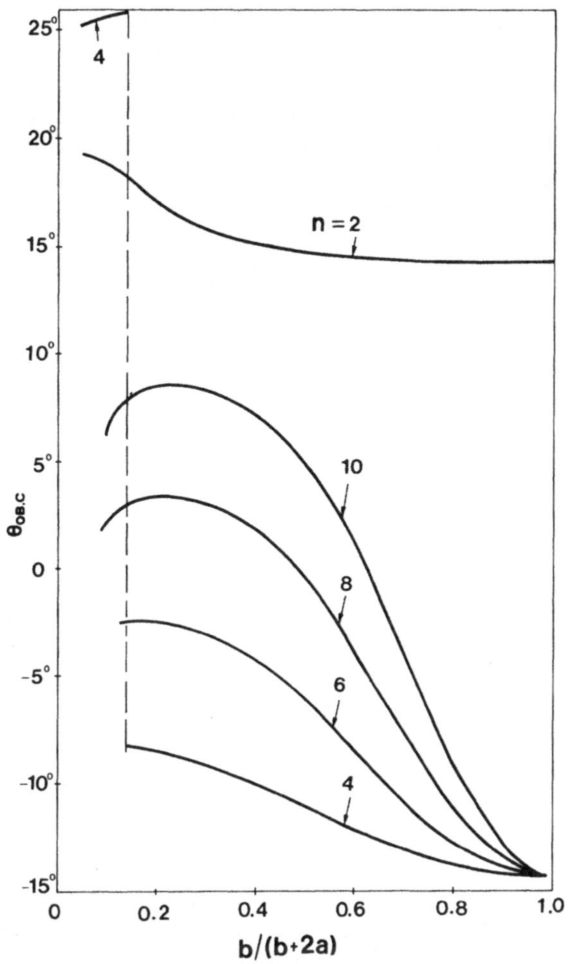

Figure 5.17. Values of the crack extension angle $\theta_{0B,C}$ for the array of cracks of Figure 5.15.

of the cracks with the tangent of the crack at this point versus $2a/b$ is shown in Figure 5.12. The values of θ_0 for $2a/b = 0$ are equal to those found in Section 5.2a for the case of a single circular crack loaded normal to its chord. It is observed that θ_0 decreases with $2a/b$. The angle formed by the fracture path with the direction of the applied stress is close to 90°. The fracture path, therefore, is approximately perpendicular to the applied stress.

Figure 5.13 illustrates the variation of the quantity $\sigma_{cr}(a/16\mu S_{cr})^{1/2}$ versus $2a/b$ for the situation when the line passing through the centers of the cracks is parallel to the applied stress. The value $2a/b = 0$ corresponds to the case of a single circular crack loaded perpendicularly to its chord. We observe that, contrary to the previous case of cracks, the critical stress for this periodic array of circular cracks is always greater than that of a single circular crack. This is due to the equal distribution of the imposed external energy between the cracks of the periodic array which lowers the stress intensity developed at the crack tips. Thus, the critical stress corresponding to the onset of crack extension is always greater for this periodic array of circular cracks than for a single circular crack. Furthermore, since the critical stress at $\beta \neq 0$ is always greater than at $\beta = 0$ (corresponding to straight cracks) the existence of curvature in the cracks always increases the critical stress. The values of the crack extension angles θ_0 for the right tips of cracks are shown in Figure 5.14. We observe that θ_0 varies slightly with $2a/b$ and, as in the previous case, (Figure 5.12) it decreases as $2a/b$ increases. Furthermore, the values of θ_0 are close to their corresponding values found for the array of cracks of Figure 5.10a. As in the latter case the fracture path is very close to the direction perpendicular to the applied stress.

5.4. A star-shaped array of circular cracks

Consider the symmetrical star-shaped array of circular cracks of Figure 5.15 consisting of n cracks. The length of each crack is equal to $2a$, the distance between the inner tips and 0 is b, and the length $2a$ corresponds to a 30° crack angle. Note that the tangent at the inner tips of the cracks passes through 0. The cracks are loaded by a constant pressure σ along their lips.

For this situation of cracks Theocaris and Ioakimidis [5] gave the values of k_1 and k_2 stress intensity factors for the inner, B, and outer, C, tips of the cracks. Using these values of k_1 and k_2 in combination with the strain energy density criterion, the critical values of the applied pressure σ for possible crack extension from the inner or the outer tips are determined. The actual critical stress σ_{cr}, which is the smaller of these two values, indicates the more vulnerable crack tip from which crack extension occurs. Figure 5.16 presents the variation of the normalized critical stress $\sigma_{cr}(a/16\mu S_{cr})^{1/2}$ versus the non-dimensional quantity $b/(b + 2a)$ for n equal to 2, 4, 6, 8 and 10. Dotted lines separate the areas in this figure in which fracture starts from the inner, B, or the outer, C, tips of the cracks, indicated by the letters, B or C in the appropriate regions. For $b/(b + 2a) \rightarrow 1$ the value of the critical stress tends to that of a single crack, and all curves for the various crack numbers n tend to the same point. It is evident from this figure that the critical stress σ_{cr} increases as the number n of cracks increases. The values $\theta_{0B,C}$ of the crack extension angle from the inner or outer tips of the cracks are shown in Figure 5.17. These angles are referred to a system of polar coordinates with origin the crack tip and the positive x-semi-axis directed outward along the crack line.

References

[1] Muskhelishvili, N.I., *Some Basic Problems of the Mathematical Theory of Elasticity*, English Translation by J.R.M. Radok, 4th Edition, Noordhoff, The Netherlands, pp. 542–545 (1953).

[2] Sih, G.C., Paris, P.C. and Erdogan, F., Crack tip stress-intensity factors for plane extension and plane bending problems, *Journal of Applied Mechanics*, 29, pp. 306–312 (1962).

[3] Gdoutos, E.E., Fracture of plates with circular cracks, *Materialprüfung*, 22, pp. 83–86 (1980).

[4] Ioakimidis, N.I. and Theocaris, P.S., Array of periodic curvilinear cracks in an infinite isotropic medium, *Acta Mechanica*, 28, pp. 239–254 (1977).

[5] Theocaris, P.S. and Ioakimidis, N.I., A star-shaped array of curvilinear cracks in an infinite isotropic elastic medium, *Journal of Applied Mechanics*, 44, pp. 619–624 (1977).

6

Cracks emanating from holes and rigid inclusions

6.1. Introduction

In the present chapter the strain energy density criterion will be used to determine the critical fracture characteristic quantities of two-dimensional bodies with cracks emanating from holes or inclusions. It has been demonstrated that brittle failure originates most frequently from stress concentration areas developed around holes or inclusions. These are frequently present in modern structures. For example, approximately one third of all cracks detected in aircrafts have been determined to originate from holes. Analysis of such failures within the framework of fracture mechanics theory is undertaken by postulating that crack-like flaws exist in the neighborhood of holes or inclusions. Thus, a great deal of research has been conducted to determine the stress field around cracks emanating from holes or inclusions. It has been found that the field in the neighborhood of the crack tip is generally governed by the values of the opening-mode and sliding-mode stress intensity factors. Studies concerning the determination of stress intensity factors at crack tips emanating from circular holes are listed in references [1–12]. Works related to the stress field around cuspidal points of rigid inclusions are listed in references [13, 14].

In this chapter we will first consider the problem of brittle failure of an isotropic homogeneous plate perforated by a circular hole with two, or an array of, equal cracks. The plate is subjected to a uniform uniaxial stress at an arbitrary inclination with respect to the crack system. Also, the general problem of fracture initiation from the boundaries of rigid inclusions which have cuspidal points and are perfectly bonded in a matrix will be studied. In all these problems, mixed mode stress conditions dominate in the vicinity of the crack tips or the cuspidal points of the inclusions. The strain energy density criterion will be used to determine the critical stress and the fracture path from the most vulnerable crack tip or cuspidal point. These results will demonstrate the dependence of the critical stress on the geometrical, mechanical and loading characteristics of the body.

6.2. Two equal cracks emanating from a circular hole

Consider an infinite isotropic elastic plate perforated by a circular hole of radius R. Two equal radial finite cracks of length l emanate from the hole (Figure 6.1). The plate is subjected to a uniform uniaxial stress σ at infinity making an angle α with the crack axes. The crack boundaries are free from external loads.

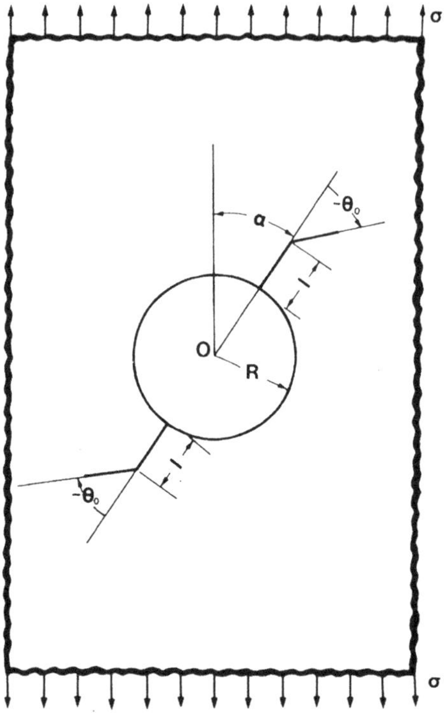

Figure 6.1. An elastic plate perforated by a circular hole with two cracks and subjected to a uniform tensile stress σ of an arbitrary inclination.

There is a critical value of a continuously increasing stress σ at which crack propagation takes place. The strain energy density criterion will be used to determine the angle θ_0 that the propagating crack forms with its initial direction, as well as the corresponding critical stress. Mixed mode conditions dominate in the vicinity of the crack tips due to the inclination of the crack axes with respect to the direction of applied stress. The values of the k_1 and k_2 stress intensity factors governing the stress field were determined by Hsu [9] for $0 \leqslant \alpha \leqslant 90^\circ$ and $0.1 \leqslant (l/R) < \infty$. For these values of k_1 and k_2, solution of equation (2.6a) of the general two-dimensional crack problem under the restrictions imposed by inequality (2.6b) gives the values θ_0 of the crack extension angle. The values σ_{cr} of the applied stress σ for crack extension are then obtained by introducing these values of θ_0 into relation (2.4) and equating the resulting minimum values S_{min} of the strain energy density factor S to its critical value S_{cr}, which is a material constant.

Figures 6.2 and 6.3 give the variation of the angle $-\theta_0$ versus the angle α of inclination of the applied stress σ with respect to the crack axis for $\lambda = l/R = 0.1$, 0.2, 0.5, 1 and ∞ when the material constant κ takes the values 1.4 (Figure 6.2) and 3.0 (Figure 6.3). All curves in Figures 6.2 and 6.3 intersect at approximately the same point for $\alpha = 30^\circ$. It is concluded, therefore, that for $\alpha = 30^\circ$ the crack extension angle is independent of the crack length. The results obtained for the values of the angle $-\theta_0$ from these figures are close to the corresponding values of this angle found

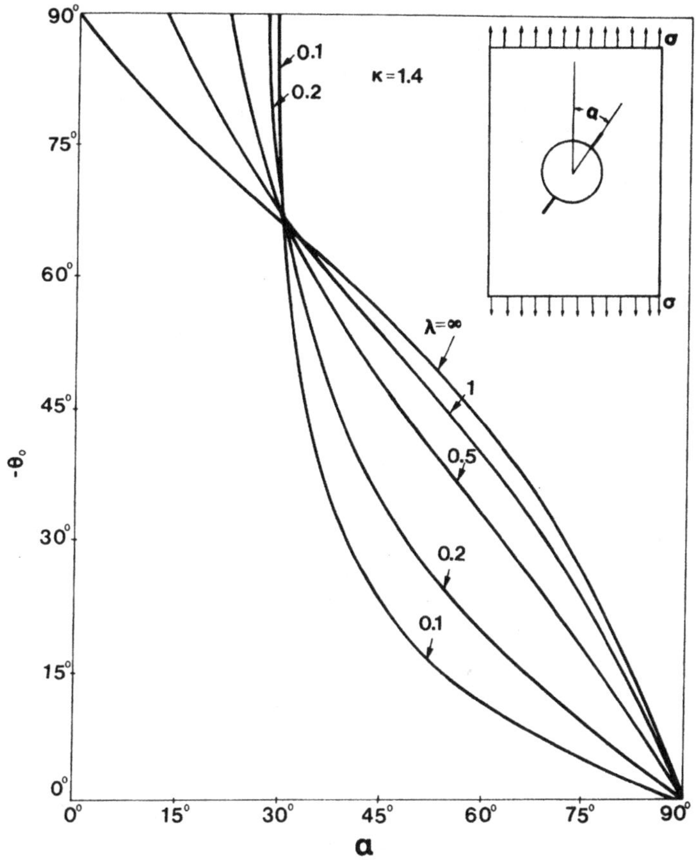

Figure 6.2. Variation of the crack extension angle $-\theta_0$ versus the angle α of the inclination of the cracks with respect to the applied stress σ for the case of Figure 6.1 with $\lambda = 0.1, 0.2, 0.5, 1.0$ and ∞. The material constant κ takes the value 1.4.

for a single crack emanating from a circular hole determined in reference [8]. The variation of the normalized critical stress $\sigma_{cr}(l/16\mu S_{cr})^{1/2}$ versus α for the cases of Figures 6.2 and 6.3 is displayed in Figures 6.4 and 6.5.

The following general conclusions concerning the dependence of the initial crack extension angle and the critical stress of fracture on the crack length, the crack inclination and the value of the Poisson's ratio of the plate are derived from the above figures:

a. *For the crack extension angle* $-\theta_0$

(i) The crack extension angle is always negative.

(ii) The angle $-\theta_0$ decreases as the crack inclination angle α with respect to the applied stress σ increases. θ_0 reaches zero when the cracks are perpendicular to σ.

(iii) For α equal to 30° the angle $-\theta_0$ is almost independent of the length of the cracks (with a deviation $\pm 1^\circ$).

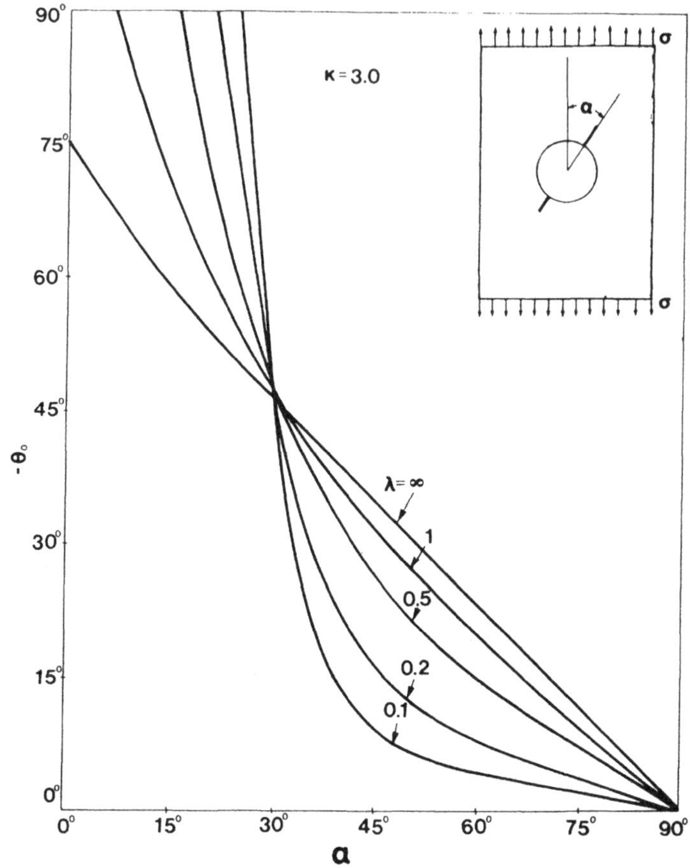

Figure 6.3. As in Figure 6.2 for $\kappa = 3.0$.

(iv) For $\alpha > 30°$ the angle $-\theta_0$ increases as the crack length increases, while quite the contrary occurs for $\alpha < 30°$.

(v) As the crack length increases above the radius of the hole for $\alpha > 30°$, the crack extension angle tends rapidly to that corresponding to an isolated crack in a plate ($\lambda \to \infty$). No interaction between the hole and the crack takes place under such conditions.

(vi) The crack extension angle $-\theta_0$ decreases as the value of the Poisson's ratio of the plate decreases.

b. *For the critical fracture stress* σ_{cr}

(i) The critical stress σ_{cr} decreases when the crack inclination angle α with respect to the applied stress increases, and reaches its minimum value when $\alpha = 90°$.

(ii) For small crack inclination angles the stress σ_{cr} decreases as the crack length increases. This holds true up to a certain angle α, after which σ_{cr} increases with the crack length.

(iii) As the crack length increases above a certain value the critical stress σ_{cr} tends

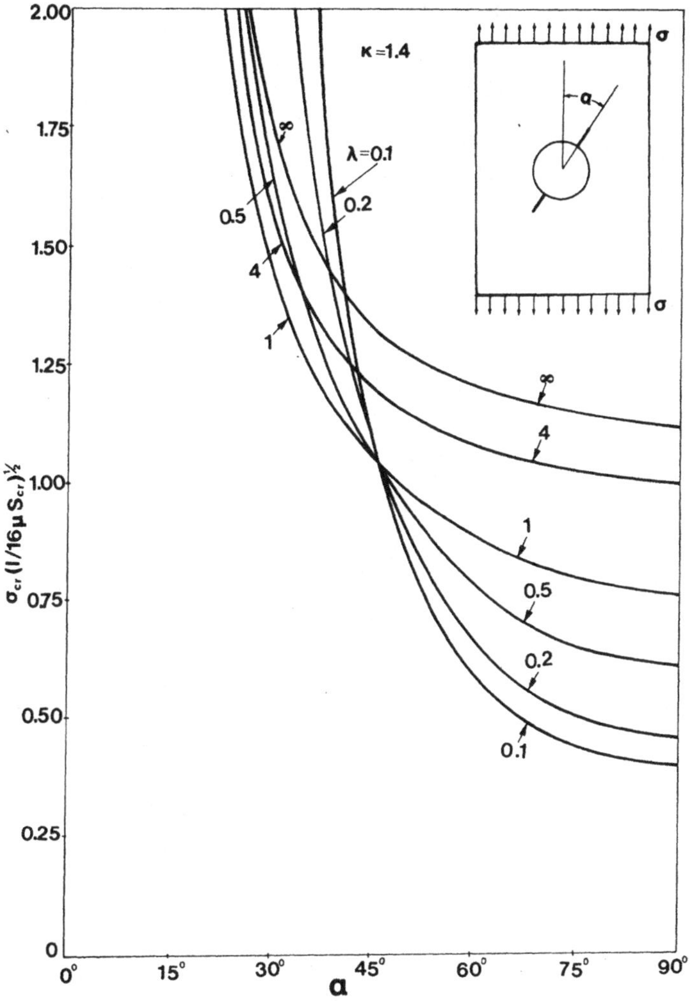

Figure 6.4. Variation of the normalized critical stress $\sigma_{cr}(l/16\mu S_{cr})^{1/2}$ versus the angle α for the case of Figure 6.1 with $\lambda = 0.1, 0.2, 0.5, 1, 4$ and ∞. The material constant κ takes the value 1.4.

to that of an isolated crack in a plate ($\lambda \to \infty$). Therefore, the interaction between the crack and the hole weakens with the increase of the crack length.

(iv) The stress σ_{cr} decreases as the value of the Poisson's ratio of the plate decreases.

For more information on the results of this section the interested reader is referred to reference [15].

6.3. An array of surface cracks emanating from a circular hole

Consider an infinite isotropic elastic plate perforated by a circular hole of radius R and a system of n symmetrically located small radial cracks of length l (Figure 6.6).

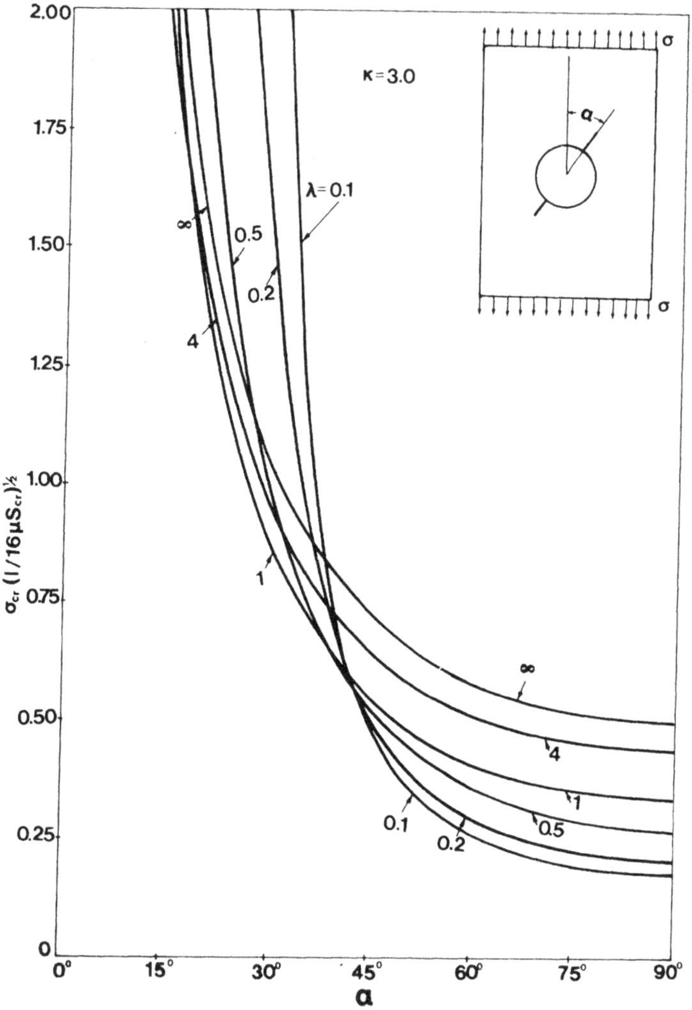

Figure 6.5. As in Figure 6.4 for $\kappa = 3.0$.

Each crack is characterized by the ordinal number j with $j = 0, 1, 2, \ldots, (n-1)$. The crack $j = 0$ is directed along the Ox-axis. The plate is subjected to a uniform uniaxial tensile stress σ at infinity forming an angle α with the Ox-axis, while the crack boundaries are free from external loads.

For this situation and assuming that the cracks are small, Kaminskii and Sailov [11] gave the following expressions for the opening- and sliding-mode $k_1^{(j)}$ and $k_2^{(j)}$ stress intensity factors at the tip of the j crack:

$$k_1^{(j)} = \frac{\sigma\sqrt{R}}{\sqrt{2}}\sqrt{\frac{1-\Delta^{2n}}{n\Delta}}\,[\tfrac{1}{2}(1+\Delta^n)]^{(2/n)-1}\,[1 - 4\cos 2(\theta_j - \alpha) -$$

$$- 2\epsilon\tilde{c}_1^n \cos(\theta_j - 2\alpha) + (A_n - \epsilon\tilde{c}_2^n)\cos 2\alpha] \tag{6.1a}$$

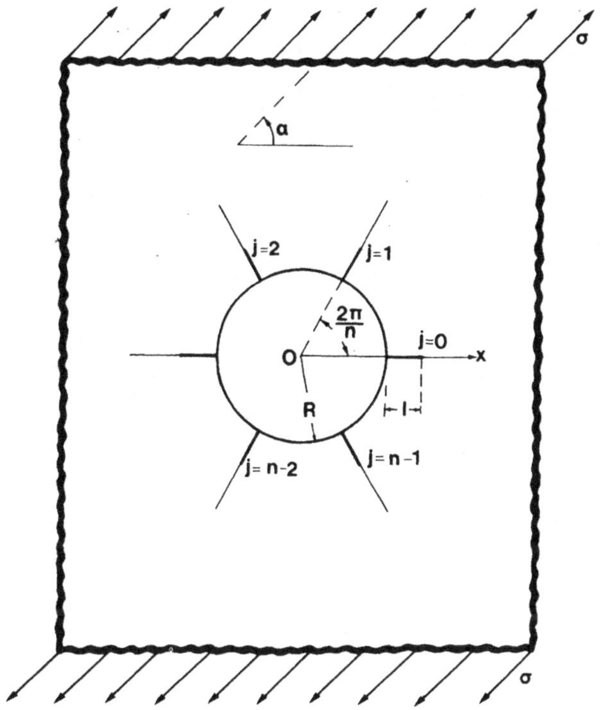

Figure 6.6. An elastic plate perforated by a circular hole with an array of n symmetrical surface cracks and subjected to an inclined uniform tensile stress σ.

$$k_2^{(j)} = \frac{\sigma\sqrt{R}}{\sqrt{2}}\sqrt{\frac{1-\Delta^{2n}}{n\Delta}}\,[\tfrac{1}{2}(1+\Delta^n)]^{(2/n)-1}\,[2\sin 2(\theta_j-\alpha) +$$

$$+ 2\epsilon\tilde{c}_1^n \sin(\theta_j - 2\alpha) + (A_n - \epsilon\tilde{c}_2^n)\sin 2\alpha] \qquad (6.1b)$$

where

$$\epsilon\tilde{c}_1^1 = \frac{2\delta^2}{(1+\Delta^{-1})^2} \qquad \epsilon\tilde{c}_2^1 = \frac{\delta^4 + 8\delta^3 + 8\delta^2}{(1+\Delta^{-1})^4}$$

$$\Delta = (1+\delta)^{-1} \qquad A_n = 2[\tfrac{1}{2}(1+\Delta^n)]^{-2/n} \qquad \delta = \frac{l}{R}$$

$$\epsilon\tilde{c}_1^2 = 0 \qquad \epsilon\tilde{c}_2^2 = \left(\frac{1-\Delta^2}{1+\Delta^2}\right)^2$$

$$\epsilon\tilde{c}_1^n = \epsilon\tilde{c}_2^n = 0 \quad (n \geqslant 3). \qquad (6.1c)$$

The critical value $\sigma_{cr}^{(j)}$ of the applied stress σ for extension of the j-crack is obtained by using the strain energy density criterion for the j-crack with the values of the stress intensity factors $k_1^{(j)}$ and $k_2^{(j)}$ given by relations (6.1). It is evident that brittle failure of the plate will take place from the extension of the crack which requires the lowest critical stress $\sigma_{cr}^{(j)}$. Therefore, the critical value σ_{cr} of the applied stress for failure of the plate is given by the following relation:

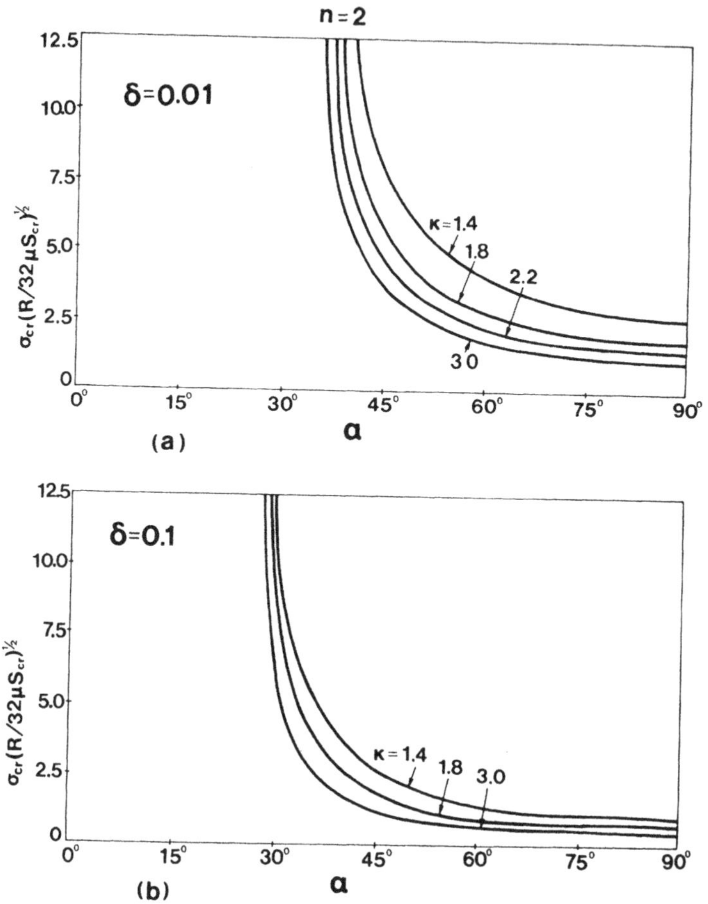

Figure 6.7. Variation of the normalized critical failure stress $\sigma_{cr}(R/32\mu S_{cr})^{1/2}$ versus the angle α of inclination of the applied stress with respect to the x-axis for the case of Figure 6.6 with $n = 2$ and $\delta = 0.01$ (a), and $\delta = 0.1$ (b). The material constant κ takes the values: 1.4, 1.8, 2.2 and 3.0.

$$\sigma_{cr} = \min\,(\sigma_{cr}^{(0)}, \sigma_{cr}^{(1)}, \ldots, \sigma_{cr}^{(n-1)}) \tag{6.2}$$

Following this procedure, the values of the critical stress σ_{cr} of the plate are determined for various geometrical configurations of the cracks. Figure 6.7 presents the variation of the quantity $\sigma_{cr}(R/32\mu S_{cr})^{1/2}$ versus the angle α of inclination of the applied stress with respect to the Ox-axis for a plate containing two radial cracks. The ratio δ of the crack length l to the hole radius R takes the values 0.01 in Figure 6.7a and 0.1 in Figure 6.7b. The material constant κ is equal to 1.4, 1.8, 2.2 and 3.0. Similar results are presented in Figures 6.8, 6.9 and 6.10 for $n = 3$, 5, and 20, respectively. These figures establish the dependence of the critical stress σ_{cr} for brittle failure of the plate caused by propagation of the more vulnerable crack j on the angle α of the inclination of the applied stress and the material constant κ. An estimate of the influence of the crack length and the number of cracks on the stress σ_{cr}

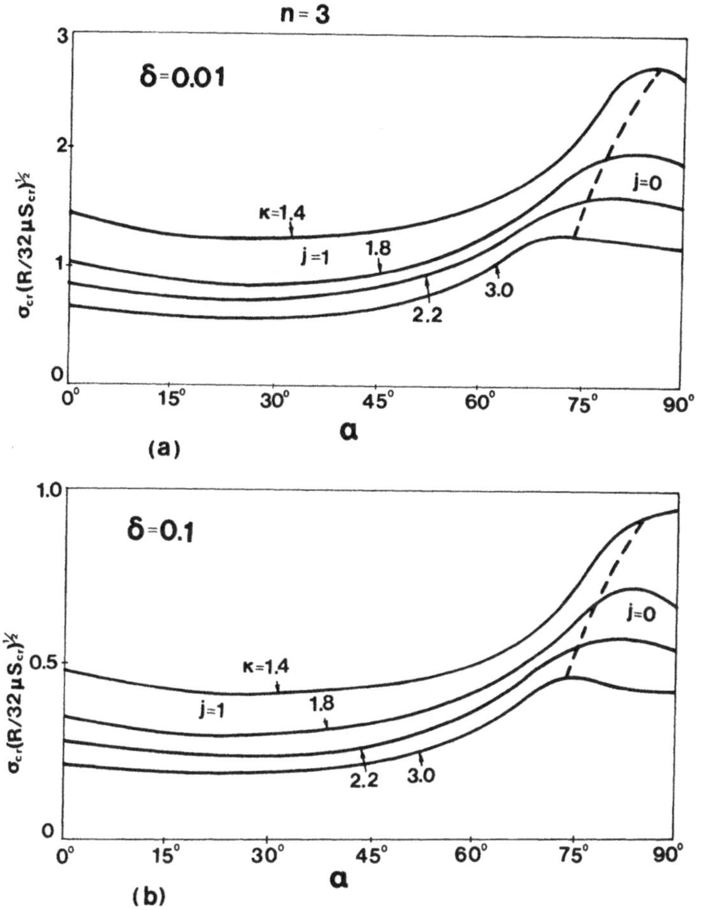

Figure 6.8. As in Figure 6.7 for $n = 3$. Dashed lines separate areas in which failure of the plate starts by propagating the cracks numbered $j = 0$ and $j = 1$.

can be obtained from Figures 6.11, 6.12, 6.13 and 6.14 which present the variation of the quantity $\sigma_{cr}(R/32\mu S_{cr})^{1/2}$ versus $\delta(0.01 < \delta < 0.10)$ for $n = 3, 4, 5, 6, 7, 8, 9,$ 10 and 20 and $\alpha = 0$, 30°, 60° and 90°. The constant κ in all these figures is set at 1.8. The values of the quantity $\sigma_{cr}(R/32\mu S_{cr})^{1/2}$ for $n = 1, 2, 3, 4, 5, 6, 7, 8, 9, 10$ and 20 given in Tables 6.1, 6.2, 6.3 and 6.4 for $\alpha = 0°, 30°, 60°$ and 90° respectively provide a clearer insight into the dependence of the critical stress σ_{cr} on the number n of the radial cracks.

The above figures and tables clearly establish the dependence of the critical stress on the number of cracks, the inclination of the applied stress, the length of the cracks and the value of the Poisson's ratio of the material of the plate. This dependence is governed by the following rules:

(i) The stress σ_{cr} decreases as the length of the cracks increases.

(ii) The stress σ_{cr} decreases as the Poisson's ratio of the plate decreases.

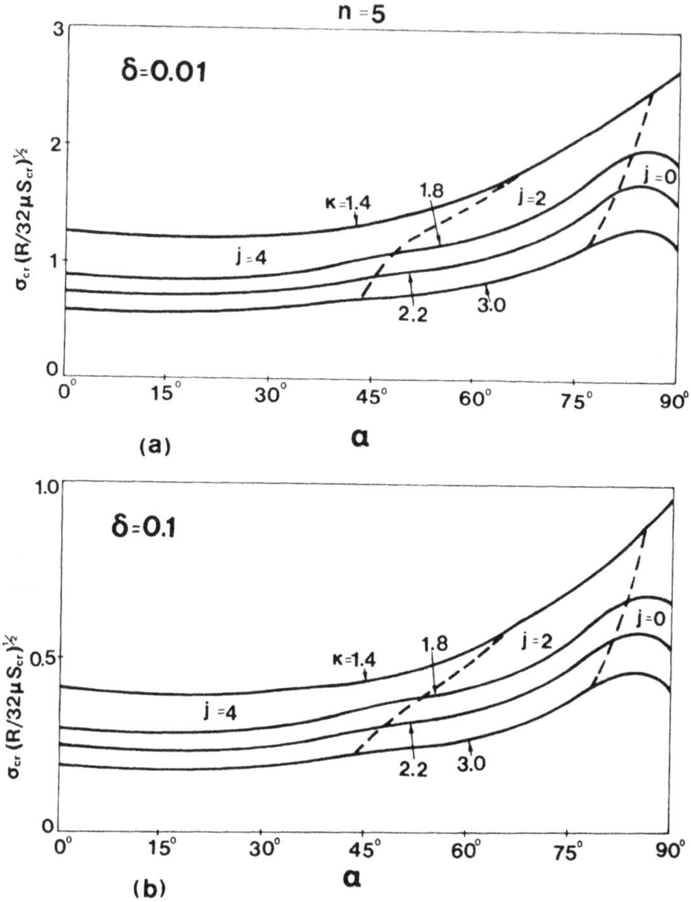

Figure 6.9. As in Figure 6.7 for $n = 5$. Dashed lines separate areas in which failure of the plate starts by propagating the cracks numbered $j = 0$, 2 and 4.

Table 6.1. Values of the normalized critical stress of fracture $\sigma_{cr}(R/32\mu S_{cr})^{1/2}$ for the case of Figure 6.6 with $\delta = 0.01, 0.04, 0.1$ and $\kappa = 1.4, 2.2, 3.0$ versus the number n of the radial cracks. The angle α is equal to 0.

	$\delta = 0.01$			$\delta = 0.04$			$\delta = 0.1$		
n	$\kappa = 1.4$	$\kappa = 2.2$	$\kappa = 3.0$	$\kappa = 1.4$	$\kappa = 2.2$	$\kappa = 3.0$	$\kappa = 1.4$	$\kappa = 2.2$	$\kappa = 3.0$
3	1.4723	0.8620	0.6711	0.7449	0.4359	0.3393	0.4139	0.2414	0.1877
4	1.1346	0.6551	0.5074	0.5750	0.3320	0.2571	0.3729	0.2153	0.1668
5	1.2482	0.7219	0.5596	0.6323	0.3657	0.2835	0.4103	0.2373	0.1839
6	1.4724	0.8620	0.6712	0.7456	0.4363	0.3397	0.4842	0.2831	0.2204
7	1.1916	0.6883	0.5333	0.6044	0.3491	0.2705	0.3942	0.2277	0.1764
8	1.1348	0.6552	0.5075	0.5761	0.3326	0.2576	0.3772	0.2178	0.1687
9	1.1690	0.6750	0.5229	0.5938	0.3429	0.2656	0.3901	0.2253	0.1745
10	1.2485	0.7221	0.5598	0.6345	0.3670	0.2845	0.4184	0.2491	0.1875
20	1.1361	0.6559	0.5081	0.5863	0.3385	0.2622	0.4114	0.2375	0.1840

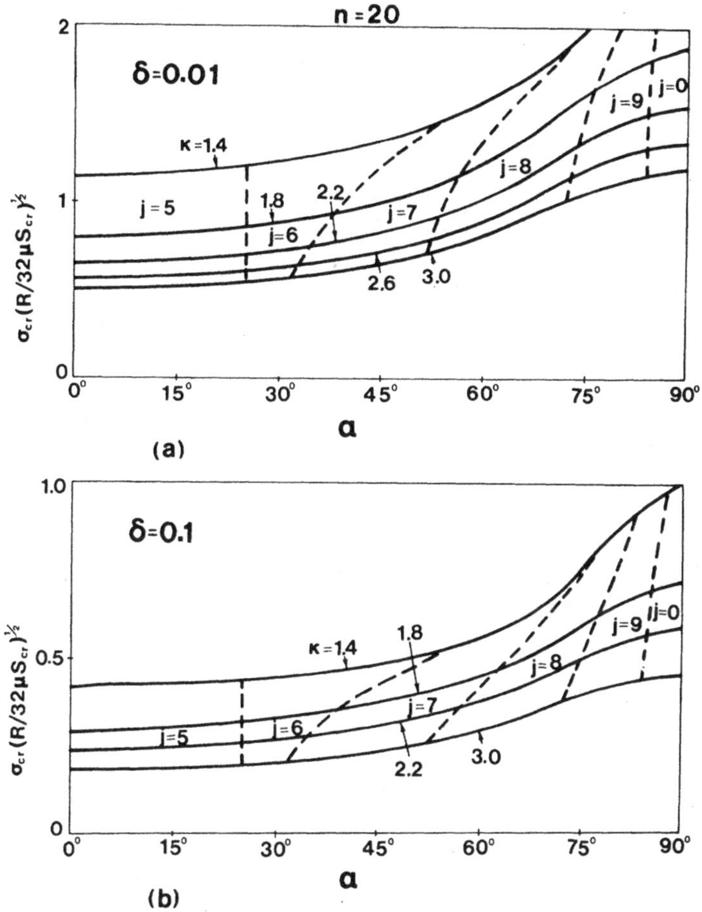

Figure 6.10. As in Figure 6.7 for $n = 20$. Dashed lines separate areas in which failure of the plate starts by propagating the cracks numbered $j = 0, 5, 6, 7, 8$ and 9.

Table 6.2. As in Table 6.1 for $\alpha = 30°$.

	$\delta = 0.01$			$\delta = 0.04$			$\delta = 0.1$		
n	$\kappa = 1.4$	$\kappa = 2.2$	$\kappa = 3.0$	$\kappa = 1.4$	$\kappa = 2.2$	$\kappa = 3.0$	$\kappa = 1.4$	$\kappa = 2.2$	$\kappa = 3.0$
1	312.8679	225.4659	188.3077	41.1663	29.6654	24.7761	11.4934	8.2824	6.9173
2	312.9023	225.4932	188.3312	41.1738	29.6708	24.7806	11.5055	8.2911	6.9246
3	1.2585	0.7325	0.5692	0.6381	0.3716	0.2889	0.4139	0.2414	0.1877
4	1.3040	0.8437	0.6781	0.6586	0.4266	0.3430	0.4246	0.2756	0.2217
5	1.2320	0.7331	0.5738	0.6240	0.3717	0.2911	0.4047	0.2416	0.1893
6	1.2586	0.7326	0.5693	0.6386	0.3720	0.2891	0.4161	0.2426	0.1886
7	1.2413	0.7511	0.5910	0.6289	0.3810	0.2999	0.4094	0.2485	0.1958
8	1.3042	0.8322	0.6448	0.6600	0.4232	0.3279	0.4299	0.2777	0.2152
9	1.2508	0.7327	0.5693	0.6344	0.3727	0.2897	0.4157	0.2454	0.1908
10	1.2322	0.7332	0.5739	0.6261	0.3730	0.2920	0.4127	0.2462	0.1929
20	1.2335	0.7340	0.5745	0.6362	0.3789	0.2967	0.4461	0.2659	0.2083

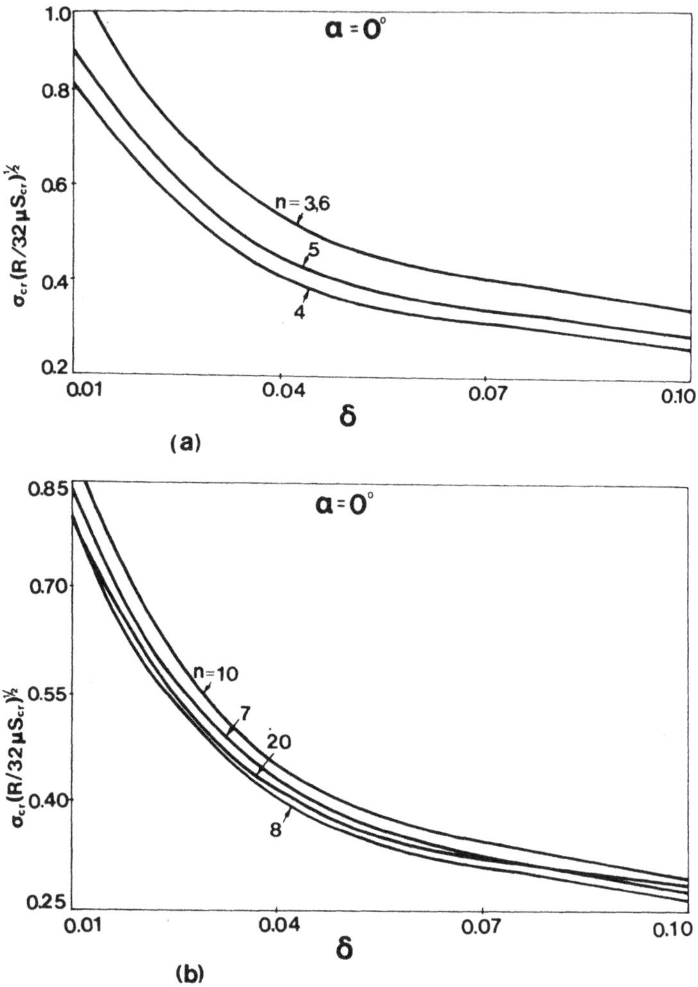

Figure 6.11. Variation of the normalized critical failure stress $\sigma_{cr}(R/32\mu S_{cr})^{1/2}$ versus the ratio δ of the crack length l to the hole radius R for the case of Figure 6.6 for $\alpha = 0$. The material constant κ is equal to 1.8, while the number n of cracks takes the values 3, 4, 5, 6 (a) and 7, 8, 10, 20 (b).

(iii) Brittle fracture of the plate occurs due to propagation of the crack which is closer to the perpendicular direction to the applied stress.

(iv) No definite rule governing the effect of the number of cracks on the stress σ_{cr} can be stated. This effect is greatly dependant on the configuration of the cracks and the inclination of the applied stress with respect to the system of cracks.

For more details on the results of this section the interested reader is referred to reference [16].

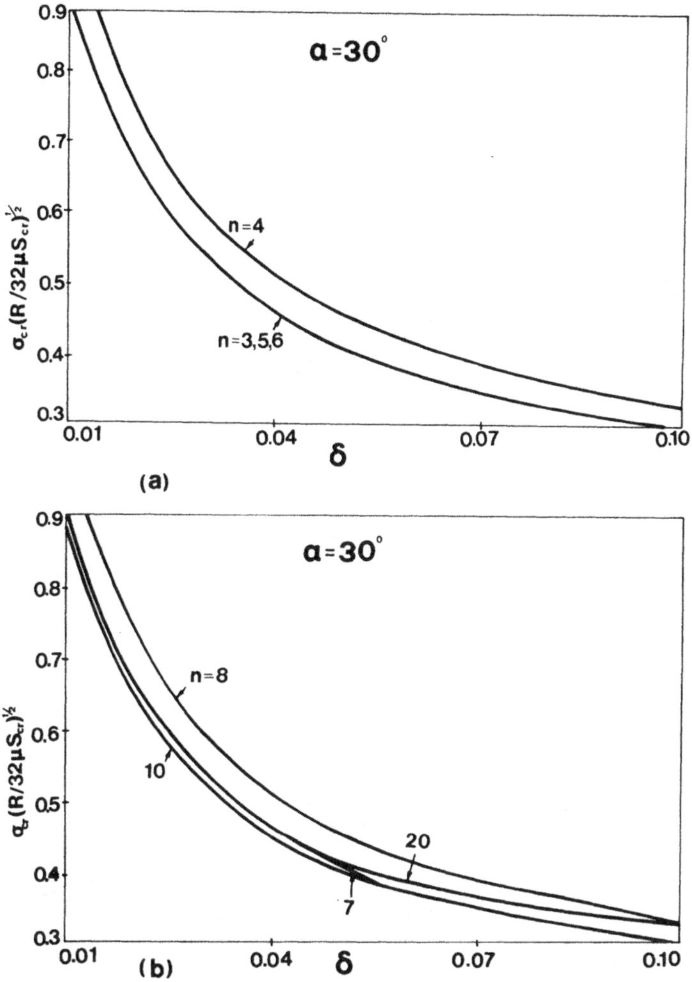

Figure 6.12. As in Figure 6.11 for $\alpha = 30°$.

Table 6.3. As in Table 6.1 for $\alpha = 60°$.

	$\delta = 0.01$			$\delta = 0.04$			$\delta = 0.1$		
n	$\kappa = 1.4$	$\kappa = 2.2$	$\kappa = 3.0$	$\kappa = 1.4$	$\kappa = 2.2$	$\kappa = 3.0$	$\kappa = 1.4$	$\kappa = 2.2$	$\kappa = 3.0$
1	4.0029	2.3111	1.7902	2.0735	1.1972	0.9273	1.3991	0.8079	0.6258
2	4.0021	2.3106	1.7898	2.0731	1.1969	0.9271	1.3974	0.8069	0.6250
3	1.5433	1.1135	0.9305	0.7802	0.5650	0.4728	0.5036	0.3671	0.3080
4	1.9645	1.7359	1.6026	0.9895	0.8744	0.8072	0.6351	0.5612	0.5181
5	1.5800	1.0163	0.8004	0.8524	0.5189	0.4091	0.5172	0.3413	0.2696
6	1.5434	1.1136	0.9305	0.7810	0.5655	0.4732	0.5066	0.3689	0.3094
7	1.5618	1.0454	0.8480	0.7920	0.5323	0.4334	0.5163	0.3494	0.2851
8	1.5947	1.0172	0.8138	0.8105	0.5191	0.4158	0.5314	0.3425	0.2750
9	1.5436	1.0107	0.8008	0.7827	0.5166	0.4098	0.5127	0.3427	0.2723
10	1.5853	1.0165	0.8005	0.8037	0.5205	0.4103	0.5279	0.3469	0.2739
20	1.5551	1.0175	0.8013	0.8018	0.5282	0.4164	0.5621	0.3730	0.2943

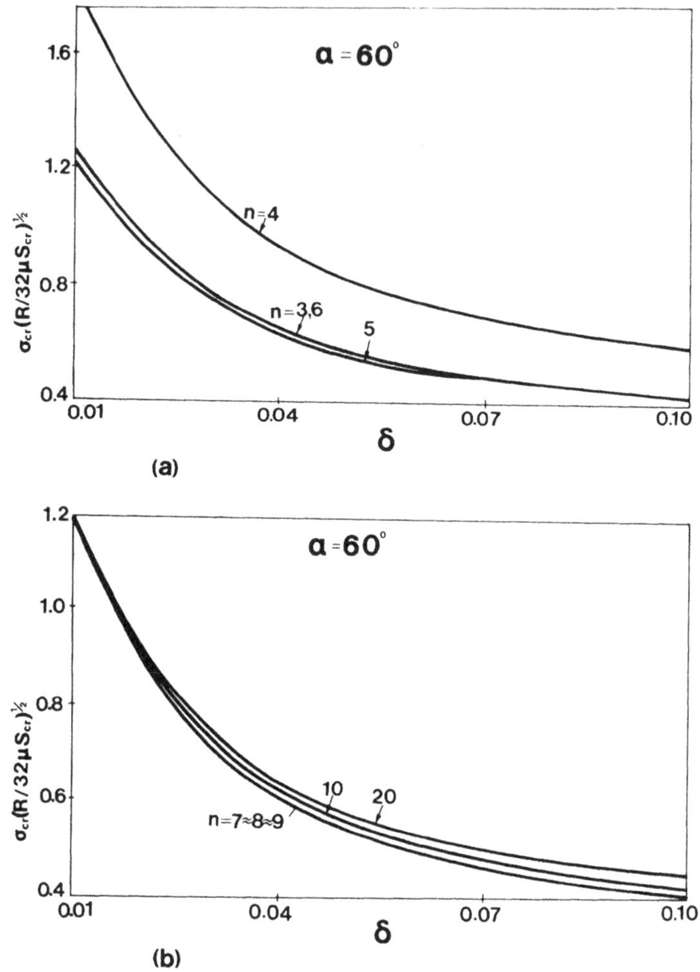

Figure 6.13. As in Figure 6.11 for $\alpha = 60°$.

Table 6.4. As in Table 6.1 for $\alpha = 90°$.

	$\delta = 0.01$			$\delta = 0.04$			$\delta = 0.1$		
n	$\kappa = 1.4$	$\kappa = 2.2$	$\kappa = 3.0$	$\kappa = 1.4$	$\kappa = 2.2$	$\kappa = 3.0$	$\kappa = 1.4$	$\kappa = 2.2$	$\kappa = 3.0$
1	2.6727	1.5432	1.1953	1.3923	0.8039	0.6227	0.9511	0.5491	0.4254
2	2.6727	1.5432	1.1953	1.3919	0.8037	0.6225	0.9499	0.5484	0.4248
3	2.6727	1.5432	1.1953	1.3921	0.8038	0.6226	0.9508	0.5490	0.4252
4	2.6727	1.5432	1.1953	1.3919	0.8037	0.6225	0.9495	0.5482	0.4247
5	2.6727	1.5432	1.1953	1.3918	0.8036	0.6224	0.9490	0.5479	0.4244
6	2.6727	1.5432	1.1953	1.3919	0.8037	0.6225	0.9491	0.5480	0.4245
7	2.6727	1.5432	1.1953	1.3921	0.8038	0.6226	0.9500	0.5484	0.4248
8	2.6727	1.5432	1.1953	1.3925	0.8040	0.6227	0.9515	0.4757	0.4255
9	2.6727	1.5432	1.1953	1.3931	0.8044	0.6231	0.9537	0.5506	0.4265
10	2.6727	1.5432	1.1953	1.3938	0.8048	0.6234	0.9565	0.5522	0.4278
20	2.6753	1.5446	1.1965	1.4109	0.8146	0.6310	1.0141	0.5855	0.4535

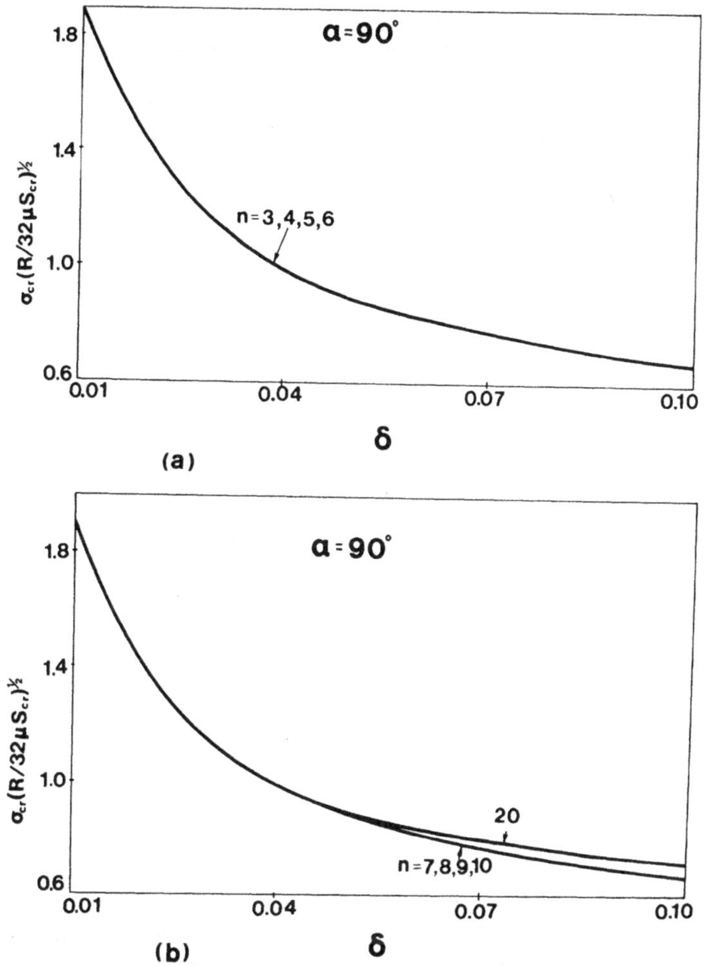

Figure 6.14. As in Figure 6.11 for $\alpha = 90°$.

6.4. Fracture of a plate with a rigid inclusion having cuspidal points

Consider a rigid inclusion with a cuspidal corner O, perfectly bonded to an infinite, isotropic elastic plate which is subjected to a system of stresses at infinity (Figure 6.15). A reference frame of Cartesian coordinates is attached to the point O with the x-axis along the tangent of the boundary of the inclusion at O. Panasyuk *et al.* [13], gave the following expressions for the polar components of stress σ_r, σ_θ, $\tau_{r\theta}$ in the vicinity of the point O

$$\sigma_r = \frac{1}{4\sqrt{2r}}\left[k_1\left[5\cos\frac{\theta}{2} + (2\kappa+1)\cos\frac{3\theta}{2}\right] - k_2\left[5\sin\frac{\theta}{2} + (2\kappa-1)\sin\frac{3\theta}{2}\right]\right] \tag{6.3a}$$

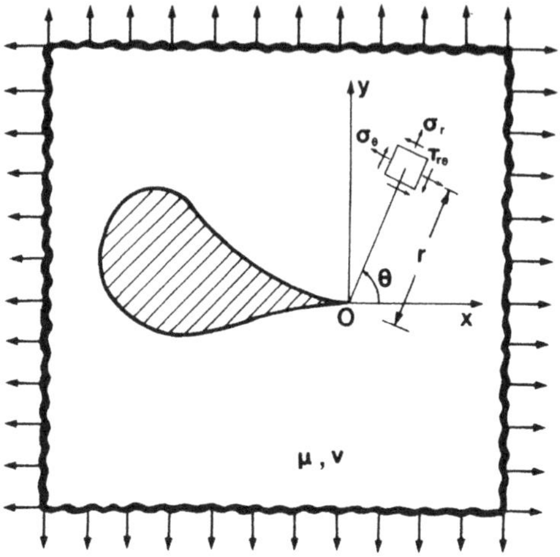

Figure 6.15. Geometry of a general shaped inclusion with a cuspidal point 0 embedded in a matrix.

$$\sigma_\theta = \frac{1}{4\sqrt{2r}}\left[k_1\left[3\cos\frac{\theta}{2} - (2\kappa+1)\cos\frac{3\theta}{2}\right] - k_2\left[3\sin\frac{\theta}{2} - (2\kappa-1)\sin\frac{3\theta}{2}\right]\right] \tag{6.3b}$$

$$\tau_{r\theta} = \frac{1}{4\sqrt{2r}}\left[k_1\left[\sin\frac{\theta}{2} - (2\kappa+1)\sin\frac{3\theta}{2}\right] + k_2\left[\cos\frac{\theta}{2} - (2\kappa-1)\cos\frac{3\theta}{2}\right]\right] \tag{6.3c}$$

where the coefficients k_1 and k_2 are called symmetrical and skew-symmetrical stress intensity factors respectively. k_1 and k_2 are independent of the coordinates r, θ and depend only on the loading conditions, the plate material and the geometrical shape of the inclusion at the cuspidal point. The coefficient κ takes the value $\kappa = (3-\nu)/(1+\nu)$ for generalized plane stress and $\kappa = 3-4\nu$ for plane strain conditions with ν representing the Poisson's ratio of the plate material.

The relations (6.3a–c) illustrate the inverse square root singularity of the stress field in the vicinity of the cuspidal point identical to that of a crack in a homogeneous elastic medium. Furthermore, for $\kappa = -1$, relations (6.3a–c) give the singular stresses in the vicinity of the tip of a crack in a mixed mode stress field (relations (2.1)). However, it should be noted that the value of $\kappa = -1$ has no physical significance, since it is $1 \leqslant \kappa \leqslant 3$ according to the above definition of κ with the values of the Poisson's ratio ν lying in the interval (0, 0.5) for isotropic materials.

The cuspidal points are the only points of the boundary of a rigid inclusion in which the stress field presents singularity and therefore high intensity stresses develop in the vicinity of these points. It follows that fracture of the composite plate would initiate from the more vulnerable cuspidal point of the boundary of the inclusion. The strain energy density criterion is used to determine the fracture path and the corresponding critical stress. The values of the stress components σ_r, σ_θ, $\tau_{r\theta}$ from relations (6.3) are introduced into relations (1.2) and (1.4) and, after lengthy calculations, the following expression for the strain energy density factor S is obtained:

$$S = a_{11}k_1^2 + 2a_{12}k_1k_2 + a_{22}k_2^2 \tag{6.4}$$

where the coefficients a_{ij} $(i, j = 1, 2)$ are given by:

$$\begin{aligned} 16\mu a_{11} &= 2(\kappa - 1)\cos^2\frac{\theta}{2} + \kappa^2 + (2\kappa + 1)\cos^2\theta \\ 16\mu a_{12} &= -[(\kappa - 1) + 2\kappa\cos\theta]\sin\theta \\ 16\mu a_{22} &= 2(\kappa - 1)\sin^2\frac{\theta}{2} + \kappa^2 - (2\kappa - 1)\cos^2\theta. \end{aligned} \tag{6.5}$$

Substituting the value of S into the basic relations (2.5) of the strain energy density criterion for the general two-dimensional crack problem the following relations are obtained:

$$\begin{aligned} &[(\kappa - 1) + 2(2\kappa + 1)\cos\theta]\sin\theta k_1^2 + 2[(\kappa - 1)\cos\theta + 2\kappa\cos 2\theta]k_1k_2 + \\ &\quad + [-(\kappa - 1) - 2(2\kappa - 1)\cos\theta]\sin\theta k_2^2 = 0 \end{aligned} \tag{6.6}$$

$$\begin{aligned} &[-(\kappa - 1)\cos\theta - 2(2\kappa + 1)\cos 2\theta]k_1^2 + 2[(\kappa - 1) + 8\kappa\cos\theta]\sin\theta k_1k_2 + \\ &\quad + [(\kappa - 1)\cos\theta + 2(2\kappa - 1)\cos 2\theta]k_2^2 > 0. \end{aligned} \tag{6.7}$$

Relations (6.6) and (6.7) represent the basic formulae of the strain energy density criterion which apply to fracture initiation from cuspidal points of rigid inclusions. Solution of equation (6.6) in conjunction with inequality (6.7) gives the value θ_0 of the angle formed by the fracture path with the tangent of the inclusion at the cuspidal point. The minimum value $S_{\min}$ of the strain energy density factor S is obtained by introducing θ_0 into relations (6.4) and (6.5). $S_{\min}$ is then equated to the critical value S_{cr} of S to obtain the critical value of a progressively increasing external stress for fracture initiation.

The above procedure will now be applied to the case of a composite plate with a hypocycloidal inclusion. Consider a rigid hypocycloidal inclusion having three cuspidal points embedded in a plate which is subjected to a uniform uniaxial stress σ. The inclusion is inscribed by a circle of radius a and is referred to a reference frame Oxy with O coinciding with the center of the circle and the singular point $j = 0$ on the x-axis. The equation of the inclusion with respect to the frame Oxy is of the form:

$$z = \frac{2a}{3}\left(\zeta + \tfrac{1}{2}\zeta^{-2}\right) \tag{6.8}$$

with $z = r\,e^{i\phi}$ and $\zeta = e^{i\theta}$.

If β is the angle that the applied stress σ subtends with the x-axis, the k_1 and k_2 stress intensity factors are given by [13]:

$$k_1^{(j)} = \frac{\sqrt{2a}}{3\kappa}\,\sigma\left[\frac{\kappa - 1}{2} + \cos\left(\frac{4\pi j}{3} - 2\beta\right)\right] \tag{6.9a}$$

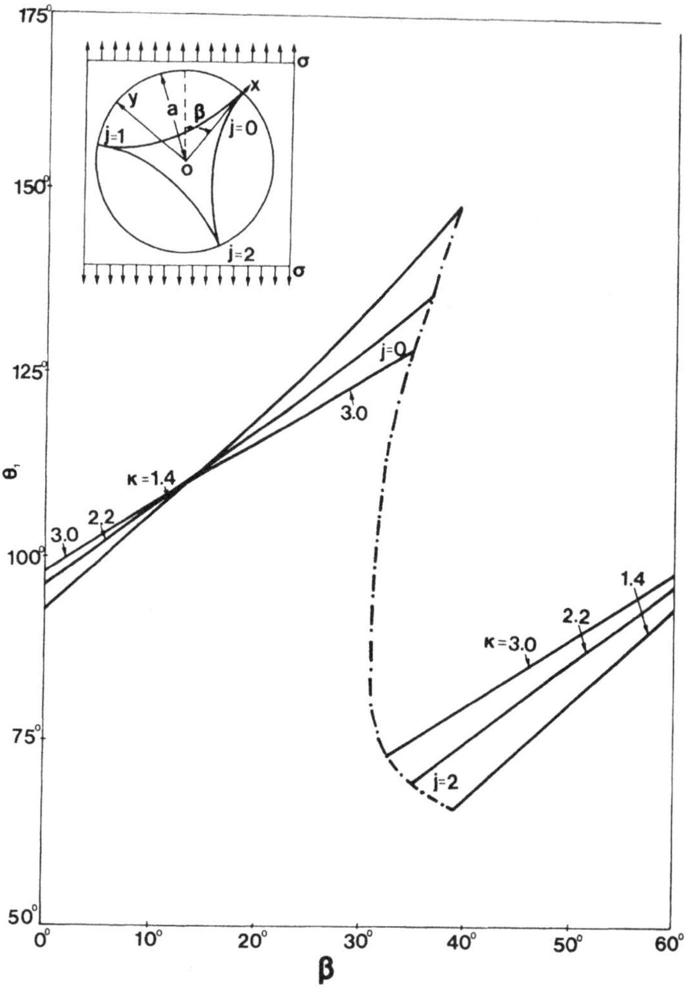

Figure 6.16. Variation of initial fracture angle θ_1 for a hypocycloidal inclusion versus the angle β of inclination of the applied tensile stress with respect to the axis Ox of the inclusion. Regions where fracture initiation starts from either of the corners $j = 0$ or $j = 2$ are separated by dotted lines.

$$k_2^{(j)} = \frac{\sqrt{2a}}{3\kappa}\,\sigma \sin\left(\frac{4\pi j}{3} - 2\beta\right) \tag{6.9b}$$

with $j = 0, 1, 2$ for the three cuspidal points of the inclusion.

An equation containing the angles θ and β and the material constant κ for each of the three cuspidal points ($j = 0$, 1, 2) of the inclusion is obtained by the introduction of the $k_1^{(j)}$ and $k_2^{(j)}$ stress intensity factors into equation (6.6). The roots θ_0 of this equation which satisfy inequality (6.7) represent the values of the angle formed by the fracture path originating from the particular cuspidal point with the tangent of the inclusion at the cuspidal point. After determining θ_0 the minimum values of the strain energy density factor are obtained. The critical values $\sigma_{cr}^{(j)}$ of the applied stress for

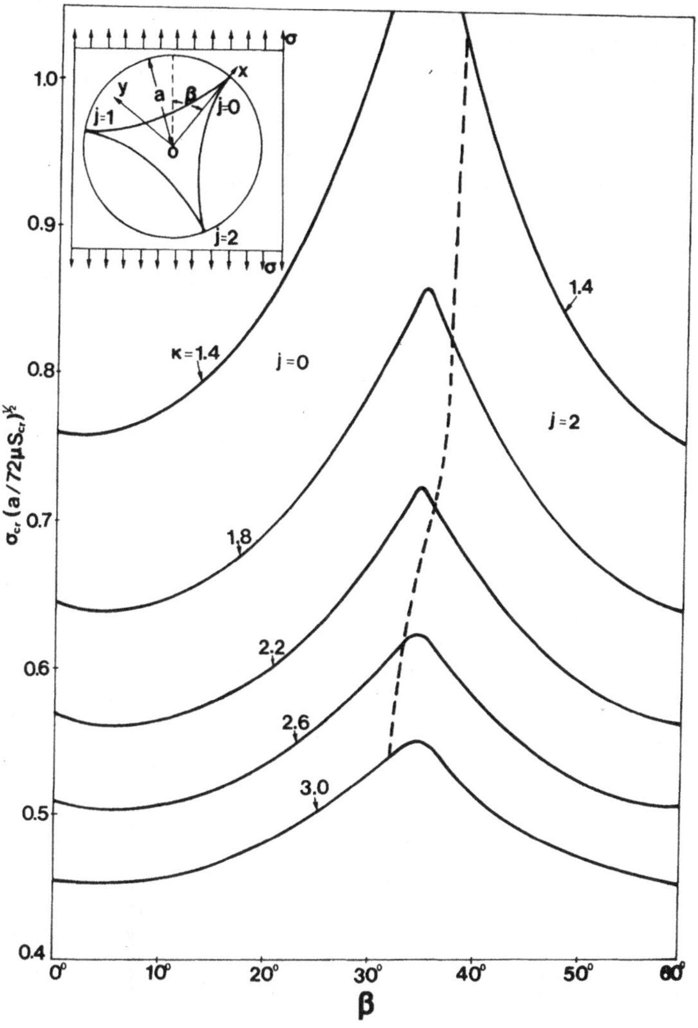

Figure 6.17. Variation of the normalized critical stress of fracture $\sigma_{cr}(a/72\mu S_{cr})^{1/2}$ for a hypocycloidal inclusion versus the angle β of inclination of the applied tensile stress with respect to the axis Ox of the inclusion. Regions where fracture initiation starts from either of the corners $j = 0$ or $j = 2$ are separated by dotted lines.

fracture initiation from each of the three cuspidal points can then be calculated. The critical stress of fracture initiation σ_{cr} equals the minimum value of the above three stresses $\sigma_{cr}^{(j)}$ ($j = 0, 1, 2$). Fracture starts from the cuspidal point with $\sigma_{cr}^{(j)} = \sigma_{cr}$. The angle β of the inclination of the inclusion with respect to the applied stress was varied in the interval $(0, 60°)$, due to the existing symmetry.

Figure 6.16 presents the variation of the fracture angle θ_1 versus β when the applied stress σ is tensile for $\kappa = 1.4$, 2.2 and 3.0. Dashed lines separate regions in which fracture takes place from the cuspidal points $j = 0$ from regions in which it takes place from $j = 2$. Figure 6.17 displays the corresponding values of the normalized fracture stress $\sigma_{cr}(a/72\mu S_{cr})^{1/2}$. Since the quantity S_{cr} is a material constant, the fracture stress σ_{cr} is indicated by this figure.

For more information on the results of this section the interested reader is referred to references [17–19].

6.5. Fracture of a plate with a rigid fiber inclusion

The stress field in the neighborhood of the end point of a rigid rectilinear inclusion embedded in an elastic plate is again given by relations (6.3) [20]. Thus the strain energy density factor S is given by relations (6.4) and (6.5), while the initial fracture angle is determined from the solution of equation (6.6). The latter must satisfy inequality (6.7). When the inclusion has a length $2l$ and the elastic plate is subjected to a uniaxial uniform stress σ at infinity making an angle β with the axis of the inclusion (Figure 6.18), the k_1 and k_2 stress intensity factors entering into relations (6.3) through (6.7) are given by [13, 20]:

$$k_1 = \frac{\sigma\sqrt{l}}{2\kappa}\left(\frac{\kappa - 1}{2} + \cos 2\beta\right) \tag{6.10a}$$

$$k_2 = -\frac{\sigma\sqrt{l}}{2\kappa}\sin 2\beta. \tag{6.10b}$$

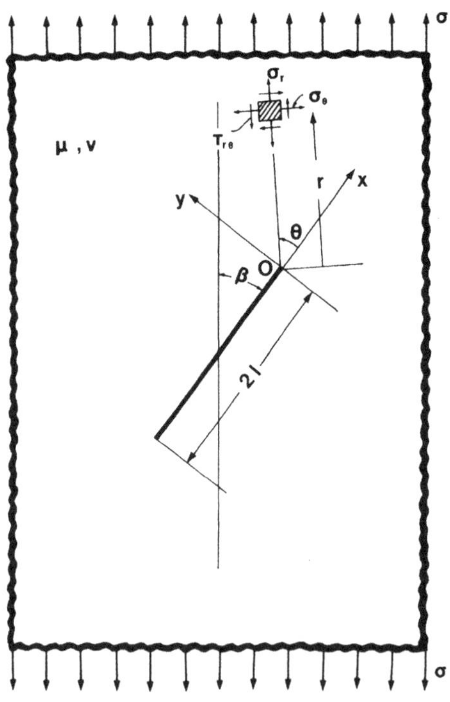

Figure 6.18. Geometry of an infinite elastic plate containing a rigid rectilinear inclusion and subjected to a uniaxial uniform stress σ.

Relations (6.10) for $\kappa = -1$ give the well-known expressions of the opening-mode k_1 and sliding-mode k_2 stress intensity factors for a crack of length $2l$ in an infinite plate subjected to a uniaxial uniform stress at an angle β with the crack axis (relations (2.7)). However, the value of $\kappa = -1$ lacks physical significance as it is noted in Section 6.4.

The values of the angle of fracture θ_0 initiating from the end points of the inclusion are determined by introducing the values of the stress intensity factors k_1, k_2 from relations (6.10) to equation (6.6). Generally four roots of equation (6.6) are obtained. Two of them must be rejected since they do not satisfy inequality (6.7), while the remaining two roots correspond to the cases of tension and compression.

For the limiting case when the axis of the fiber is parallel to the direction of the applied stress ($\beta = 0$) relations (6.10) yield the following values of the coefficients k_1 and k_2:

$$k_1 = \frac{\sigma\sqrt{l}}{2}\frac{\kappa+1}{2\kappa} \tag{6.11a}$$

$$k_2 = 0. \tag{6.11b}$$

For $k_2 = 0$ equation (6.6) gives the following values for the fracture angle θ_0

$$\theta_0 = 0 \tag{6.12a}$$

or

$$\theta_0 = \pm \arccos\left[\frac{1-\kappa}{2(2\kappa+1)}\right]. \tag{6.12b}$$

From these two values of θ the value $\theta = 0$ is rejected since it does not satisfy inequality (6.7). Thus, the fracture angle θ_0 is given by relation (6.12b). The critical stress of fracture $\sigma_{1,2cr}$ is given by:

$$\sigma_{1,2cr} = 2\sqrt{\frac{16\mu S_{cr}}{l}}\,\frac{2\kappa}{(\kappa+1)\sqrt{2(\kappa-1)\cos^2\dfrac{\theta}{2}+\kappa^2+(2\kappa+1)\cos^2\theta}} \tag{6.13}$$

(where 1 corresponds to the case of tensile stress and 2 to compressive applied stress obtained for this value of θ).

For the other limiting case when the axis of the inclusion is perpendicular to the direction of the applied stress ($\beta = 90°$), the k_1 and k_2 stress intensity factors are given by:

$$k_1 = \frac{\sigma\sqrt{l}}{2}\frac{\kappa-3}{2\kappa} \tag{6.14a}$$

$$k_2 = 0. \tag{6.14b}$$

Thus, the values of the crack extension angle θ_0 are given again by relation (6.12b), while the critical fracture stress is given by:

$$\sigma_{1,2cr} = 2\sqrt{\frac{16\mu S_{cr}}{l}}\,\frac{2\kappa}{(\kappa-3)\sqrt{2(\kappa-1)\cos^2\dfrac{\theta}{2}+\kappa^2+(2\kappa+1)\cos^2\theta}}. \tag{6.15}$$

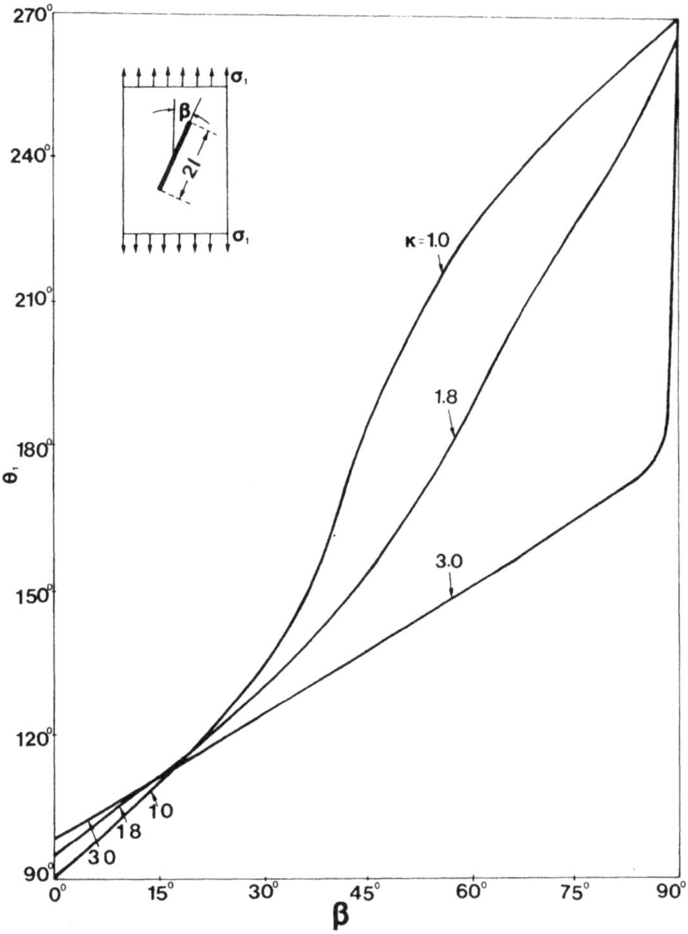

Figure 6.19. Variation of the initial fracture angle θ_1 for the case of Figure 6.18 versus the angle β of the orientation of the fiber for tensile applied loads.

Figures 6.19 and 6.20 present the variation of the fracture angles θ_1 and θ_2 for tensile and compressive loads respectively, versus the angle β of the orientation of the fiber for $\kappa = 1.0$, 1.4, 1.8, 2.2, 2.6 and 3.0. These figures illustrate the marked dependence of the angles $\theta_{1,2}$ on the value of κ, that is, on the value of the Poisson's ratio ν of the matrix material. These figures also show that θ_1 increases as angle β increases. Examining the dependence of θ_1 on the value of κ, we see that for small values of the angle β (from $\beta = 0°$ to about $\beta = 20°$) θ_1 increases with κ; while for the remaining values of β this rule is reversed and θ_1 decreases as κ increases. Figure 6.20 presents the variation of the fracture angle θ_2 under compressive applied loads and shows that θ_2 increases with β and decreases with κ.

From Figures 6.19 and 6.20 we conclude that for both tensile and compressive loads the angles θ_1 and θ_2 vary in an interval of approximately 180°. For the extreme cases of a fiber oriented along, or perpendicular to, the direction of the applied load

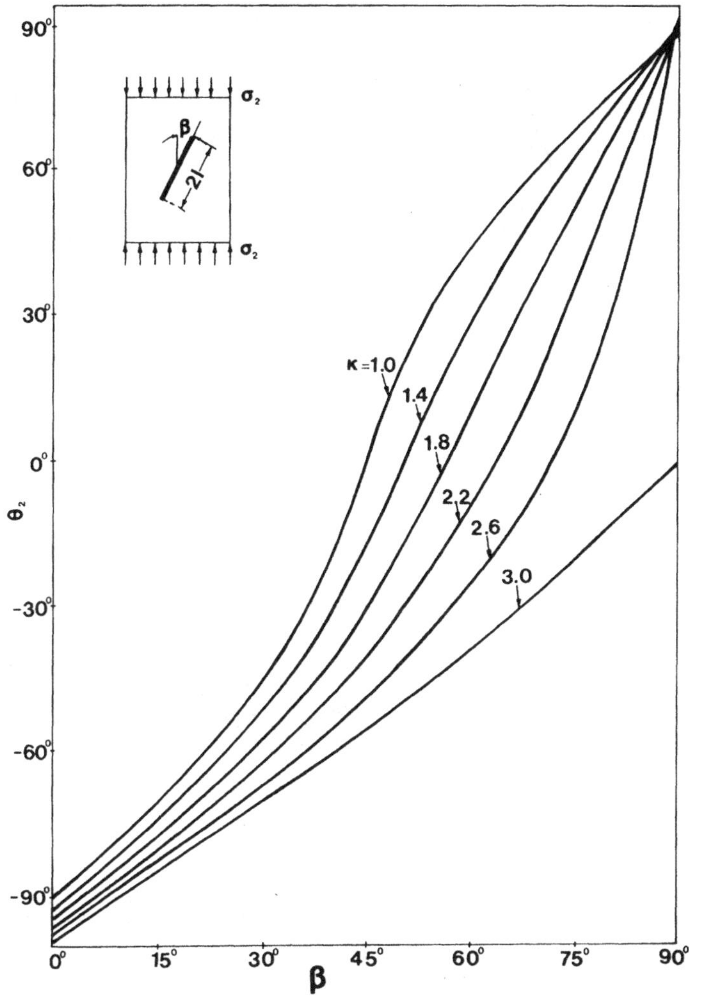

Figure 6.20. As in Figure 6.19 for compressive applied loads.

the fracture path is almost normal to the fiber. The direction of the fracture path under tensile loads is opposite that which takes place under compressive loads. When $\kappa = 1.0$, fracture path is exactly perpendicular to the fiber.

The variation of the quantities $\sigma_{1,2\mathrm{cr}}(l/64\mu S_{\mathrm{cr}})^{1/2}$ versus β for $\kappa = 1.0, 1.4, 1.8, 2.2, 2.6$ and 3.0 is presented in Figures 6.21 and 6.22. From Figure 6.21 corresponding to tensile loads, we observe that $\sigma_{1\mathrm{cr}}$ depends significantly on the value of κ. Thus, for $\kappa = 1.0$, $\sigma_{1\mathrm{cr}}$ is symmetrical with respect to the value of β equal to 45° at which $\sigma_{1\mathrm{cr}}$ becomes infinite. No such symmetries exist for the other values of κ. We see that for $\kappa = 1.4$ and 1.8, $\sigma_{1\mathrm{cr}}$ presents a maximum in the interval $0 < \beta < 90^\circ$ while for $\kappa = 2.2, 2.6$ and 3.0, $\sigma_{1\mathrm{cr}}$ takes its maximum values for $\beta = 90^\circ$. For $\kappa = 3.0$, σ_{cr} tends to infinity as β tends to 90°. Examining the dependence of $\sigma_{1\mathrm{cr}}$ on κ we observe that $\sigma_{1\mathrm{cr}}$

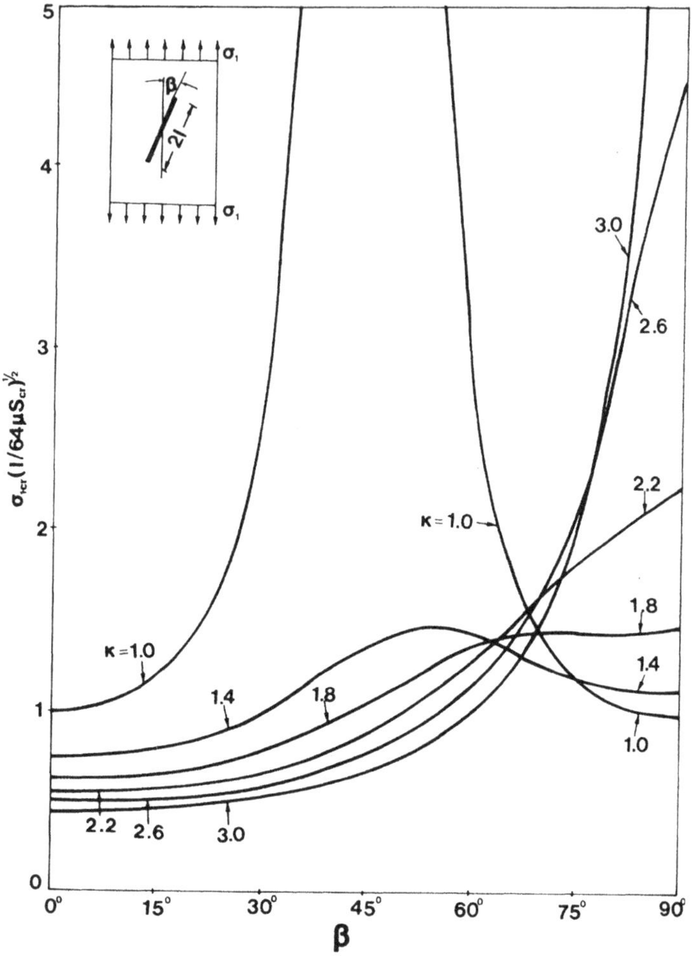

Figure 6.21. Values of the normalized critical fracture stress $\sigma_{1cr}(l/64\mu S_{cr})^{1/2}$ for the case of Figure 6.19 corresponding to tensile applied loads.

decreases as κ increases for the major part of the interval $0 < \beta < 90°$, while this rule is reversed near the end of this interval.

Similar observations can be made from a plate subjected to compressive loads (shown in Figure 6.22). Again, the curve corresponding to $\kappa = 1.0$ is symmetrical with respect to $\beta = 45°$ while the curves for $\kappa = 1.4$, 1.8 and 2.2 are bell-shaped, presenting maximum values in the interval $0 < \beta < 90°$. As with tensile loads, σ_{2cr} tends to infinity as β tends to $90°$ for $\kappa = 3$. Comparing the values of the critical stress of fracture σ_{1cr} and σ_{2cr} for tensile and compressive loads, we observe that σ_{2cr} is always greater than σ_{1cr} except when $\kappa = 1.0$, when they coincide. We also observe that for the extreme value of $\beta = 0°$ and $90°$, σ_{1cr} and σ_{2cr} coincide.

For more information on the results of this section the interested reader is referred to references [19] and [21].

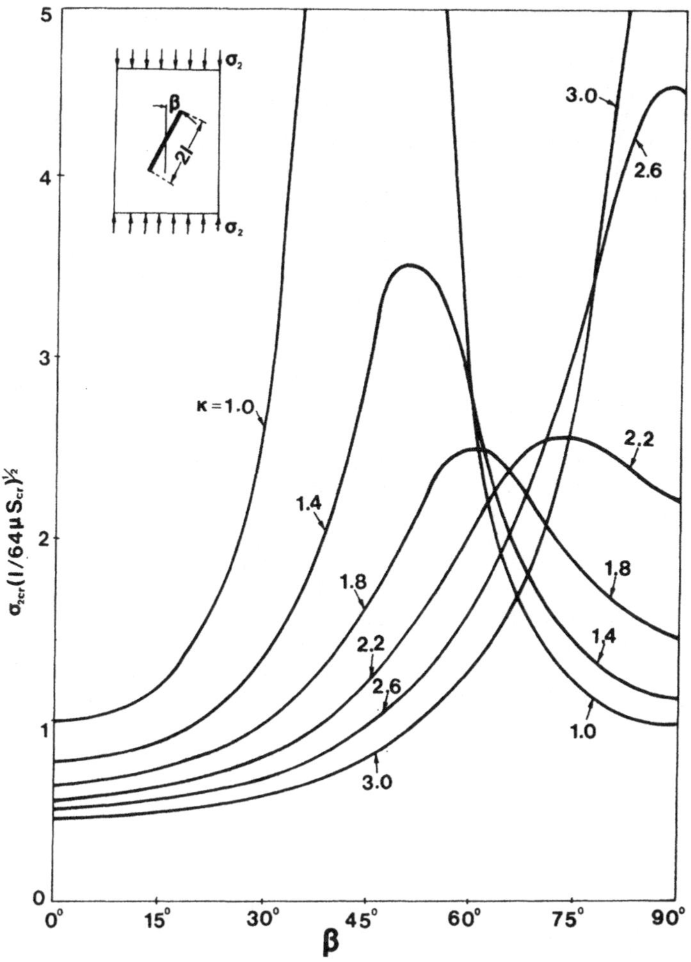

Figure 6.22. As in Figure 6.21 for compressive applied loads.

References

[1] Bowie, O.L., Analysis of an infinite plate containing radial cracks originating at the boundary of an internal circular hole, *Journal of Mathematical Physics*, 35, pp. 60–71 (1956).

[2] Paris, P.C. and Sih, G.C., Stress analysis of cracks, *Fracture Toughness Testing and its Applications*, ASTM STP-381, pp. 30–83 (1965).

[3] Roberts, R. and Rich, T., Stress intensity factors for an infinite plate with radial cracks emanating from an internal hole and subjected to cylindrical bending, *Progress Report NASA NGR-39-007-011*, Lehigh University, April (1966).

[4] Kutter, H.K., Stress analysis of a pressurized circular hole with radial cracks in an infinite elastic plate, *International Journal of Fracture Mechanics*, 6, pp. 233–247 (1970).

[5] Tweed, J. and Rooke, D.P., The distribution of stress near the tip of a radial crack at the edge of a circular hole, *International Journal of Engineering Science*, 11, pp. 1185–1195 (1973).

[6] Grand Jr., A.F., Stress intensity factors for some through-cracked fastener holes, *International Journal of Fracture*, 11, pp. 283–294 (1975).

[7] Hsu, Y.C., The infinite sheet with cracked cylindrical hole under inclined tension or in-plane shear, *International Journal of Fracture*, 11, pp. 571–581 (1975).

[8] Hsu, Y.C. and Forman, R.G., Fracture angle and strain-energy-density-factor of a crack at hole at an arbitrary angle, *Engineering Fracture Mechanics*, 7, pp. 743–750 (1975).

[9] Hsu, Y.C., The infinite sheet with two radial cracks from cylindrical hole under inclined tension or in-plane shear, *International Journal of Fracture*, 13, pp. 839–845 (1977).

[10] Karlsson, A. and Bäcklung, J., Summary of SIF design graphs for cracks emanating from circular holes, *International Journal of Fracture*, 14, pp. 585–596 (1978).

[11] Kaminskii, A.A. and Sailov, N.S., Failure of a brittle body close to a hole as a result of development of a system of surface cracks, *Problemy Prochnosti*, 10, pp. 71–76 (1973).

[12] Kaminskii, A.A., A study of the stress field near small radial cracks emanating on the contour of a hole, *Prikladnaya Mekhanika*, 7, pp. 112–115 (1971).

[13] Panasyuk, V.V., Berezhnitskii, L.T. and Trush, I.I., Stress distribution about defects such as rigid sharp-angled inclusions, *Problemy Prochnosti*, 7, pp. 3–9 (1972).

[14] Theocaris, P.S. and Paipetis, S.A., Constrained zones at singular points of inclusion contours, *International Journal of Mechanical Sciences*, 18, pp. 581–587 (1976).

[15] Gdoutos, E.E., Brittle fracture of a plate with two cracks emanating from a circular hole under inclined tension, *Materialprüfung*, 23, pp. 194–196 (1981).

[16] Gdoutos, E.E., Failure of a plate with a circular hole resulting from an array of surface cracks, *Engineering Fracture Mechanics*, 15, pp. 457–467 (1981).

[17] Gdoutos, E.E., Fracture initiation from singular points of rigid inclusions, *Journal of Applied Mechanics*, 47, Transactions ASME 102, pp. 971–973 (1980).

[18] Gdoutos, E.E., Fracture of composites with rigid inclusions having cuspidal points, *Proceedings of the International Conference on Analytical and Experimental Fracture Mechanics*, Rome, Italy, June 23–27, 1980, Editors G.C. Sih and M. Mirabile, Sijthoff & Noordhoff, pp. 943–958 (1981).

[19] Gdoutos, E.E., Fracture phenomena in composites with rigid inclusions, *Proceedings of the US-Greece Seminar on Mixed Mode Crack Propagation*, Athens, Greece, August 18–22, 1980, Editors G.C. Sih and P.S. Theocaris, Sijthoff & Noordhoff, pp. 109–122 (1981).

[20] Sih, G.C., Plane extension of rigidly embedded line inclusions, *Proceedings of the Ninth Midwestern Mechanics Conference*, Madison, Wisconsin, August 16–18, 1965, pp. 61–79 (1965).

[21] Gdoutos, E.E., Failure of a composite with a rigid fiber inclusion, *Acta Mechanica*, 39, pp. 251–262 (1981).

7

Composite materials

7.1. Introduction

Composite materials are gaining popularity in engineering applications due to their flexibility in allowing achievement of the desired mechanical and other physical properties in combination with light weight components. Two different approaches are generally used to determine the strength of composite materials. In the first, it is assumed that the material is macroscopically homogeneous (either isotropic or anisotropic). Its bulk strength is then estimated by considering a random distribution of the existing imperfections and applying statistical strength theories. The second approach focuses on the microstructure of the material. Fracture is assumed to initiate from localized imperfections, usually idealized by cracks, voids, and inclusions.

In the present chapter, the problem of brittle fracture of some types of composite materials will be studied within the framework of the latter approach by assuming that fracture starts from crack-like defects. Problems of composite materials with cracks have been studied extensively in the literature [1–36]. In these studies attention was focused primarily on local stress distribution in the neighborhood of the crack tips, governed by the values of the opening-mode and sliding-mode stress intensity factors.

This chapter aims to study the characteristics of brittle fracture of composite materials using the strain energy density criterion to determine the values of applied loads for extension of existing cracks from their more vulnerable tips, and the fracture paths. These determinations reveal the dependence of the critical fracture loads and fracture paths on the geometrical and mechanical properties of the composite and on the location of the existing cracks with respect to interfaces or inclusions.

7.2. A bimaterial plate with a crack arbitrarily oriented to the interface

Consider a two-dimensional bimaterial composite made up of two bonded half-planes, numbered 1 and 2, with elastic constants μ_1, ν_1 and μ_2, ν_2 respectively (Figure 7.1). A crack AB of length 2 inclined to the bimaterial interface under angle β is embedded in the material 1. Let δ be the distance from the crack tip B to the interface. The plate is loaded by a uniform, normal stress σ applied along the crack surfaces. This problem was considered by Ashbaugh [18] who gave the values of the k_1 and k_2 stress intensity factors for both crack tips A and B for many material combinations of the composite plate. His solution is given in terms of the bimaterial constants p and q defined by:

$$p = \frac{\mu_1/\mu_2 - 1}{\mu_1/\mu_2 + \kappa_1} \qquad q = 1 - \frac{\kappa_1 + 1}{(\mu_1/\mu_2)\kappa_2 + 1} \tag{7.1}$$

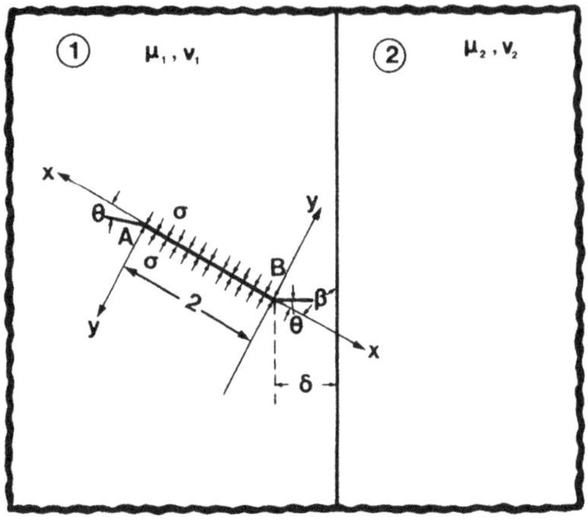

Figure 7.1. Geometry of a bimaterial plate with an arbitrarily oriented crack in one of its constituents.

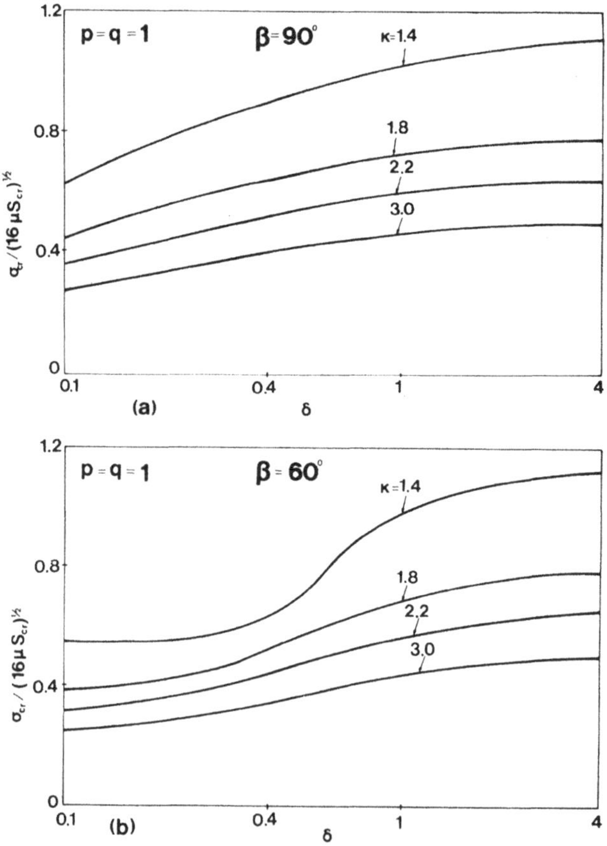

Figure 7.2. Variation of the quantity $\sigma_{cr}/(16\mu S_{cr})^{1/2}$ versus the distance δ for a cracked plate with a stress-free boundary ($p = q = 1$). The crack is loaded by a uniform compressive stress along its surfaces and is inclined at an angle β equal to 90° (a) and 60° (b) with respect to the boundary.

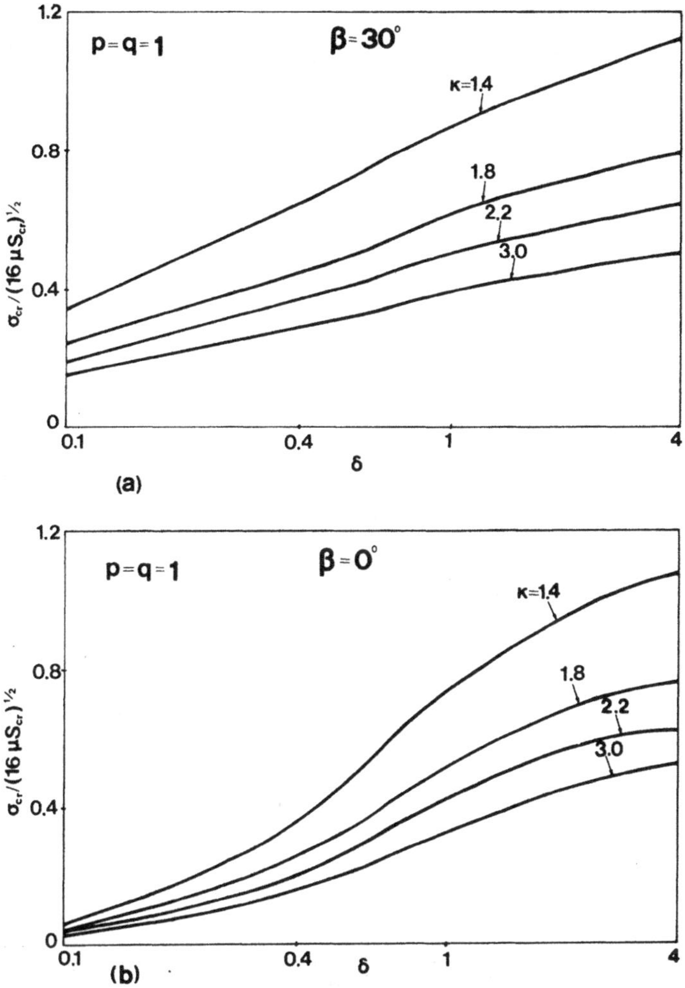

Figure 7.3. As in Figure 7.2 for the angles $\beta = 30°$ (a) and $0°$ (b).

where $\kappa_{1,2} = 3 - 4\nu_{1,2}$ for plane strain conditions, or $\kappa_{1,2} = (3 - \nu_{1,2})/(1 + \nu_{1,2})$ for generalized plane stress conditions.

The values of k_1 and k_2 are introduced into the equations of the strain energy density criterion (Section 2.2, page 11) to determine the critical stresses σ_{cr}^A and σ_{cr}^B corresponding to the onset of crack extension from its tips A and B. The critical stress σ_{cr} of the plate corresponds to the minimum value of σ_{cr}^A and σ_{cr}^B, indicating also the tip (A or B) from which crack propagation starts. Figures 7.2 and 7.3 present the variation of the quantity $\sigma_{cr}/(16\mu S_{cr})^{1/2}$ versus the distance δ of the crack tip from the interface for a single plate with a stress-free boundary ($p = q = 1$) and for the values of the angle of inclination β of the crack with respect to the interface equal to 90°, 60°, 30° and 0°. The stress σ applied along the crack surfaces is compressive. The material constant κ is assigned the values 1.4, 1.8, 2.2 and 3.0. For all cases of Figures

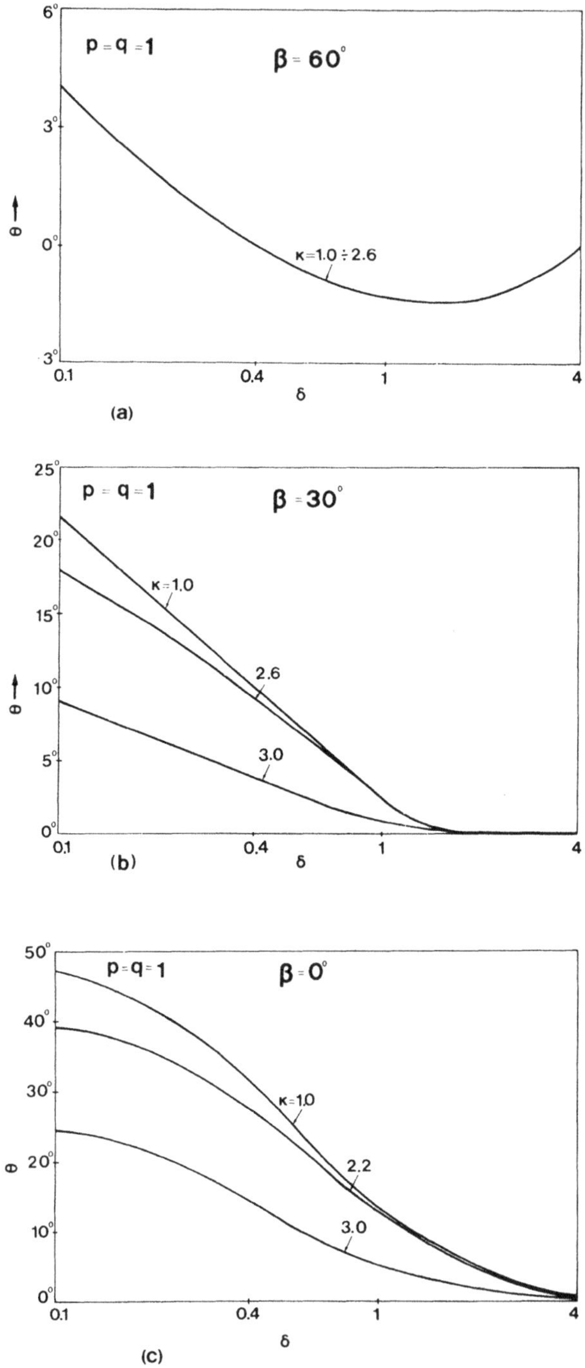

Figure 7.4. Variation of the crack extension angle θ from the tip B versus the distance δ for the cases of Figures 7.2b and 7.3.

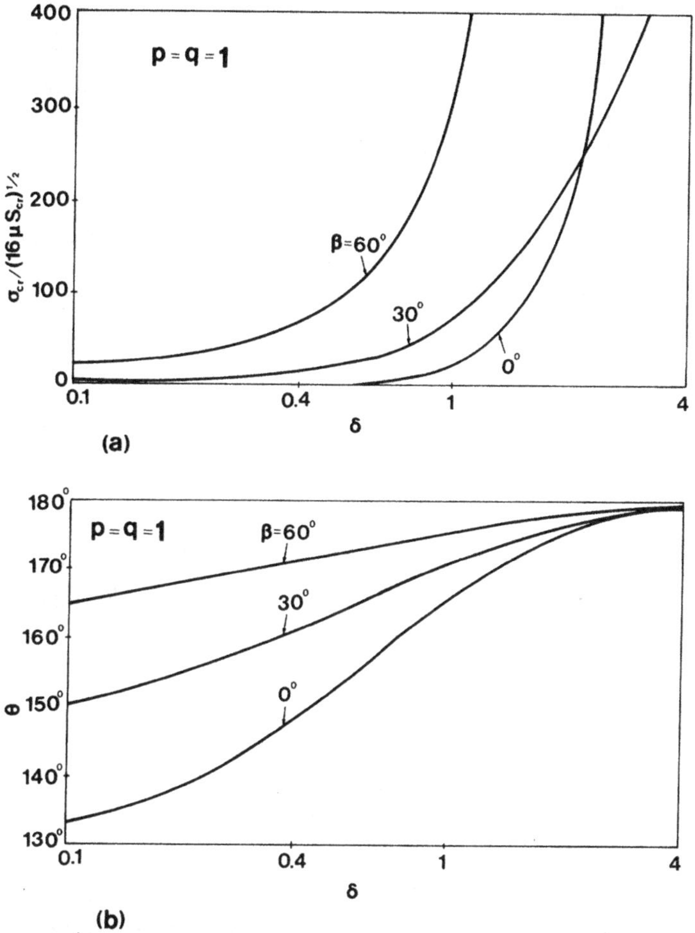

Figure 7.5. Variation of the quantity $\sigma_{cr}/(16\mu S_{cr})^{1/2}$ (a) and the crack extension angle θ (b) versus the distance δ for the case of Figure 7.1 and for the values of the crack inclination angle β with respect to the boundary of the plate equal to $0°$, $30°$, $60°$. The crack is loaded by a uniform tensile stress along its surfaces.

7.2 and 7.3 the crack propagates from tip B, which is close to the boundary. The values of the crack extension angle θ for Figures 7.2b and 7.3 are displayed in Figure 7.4. The crack in Figure 2a is perpendicular to the boundary and therefore it extends along its own plane ($\theta = 0$). The quantities $\sigma_{cr}/(16\mu S_{cr})^{1/2}$ and θ for the composite material under a tensile stress σ are displayed in Figures 7.5a and 7.5b respectively. These quantities are basically independent of the material constant κ.

Consider now a cracked plate with a rigid boundary ($q = (1/p) = -\kappa_1$). Setting $\kappa_1 = 2$, the variation of the quantities $\sigma_{cr}/(16\mu S_{cr})^{1/2}$ and θ versus δ is presented in Figures 7.6a and 7.6b for $\beta = 90°, 60°, 30°$ and 0. Contrary to the previous case of the stress-free boundary the crack always extends from tip A, located at a greater distance from the rigid boundary than tip B. Figures 7.7 and 7.8 present the variation of the quantity $\sigma_{cr}/(16\mu S_{cr})^{1/2}$ versus the crack inclination angle β for the following

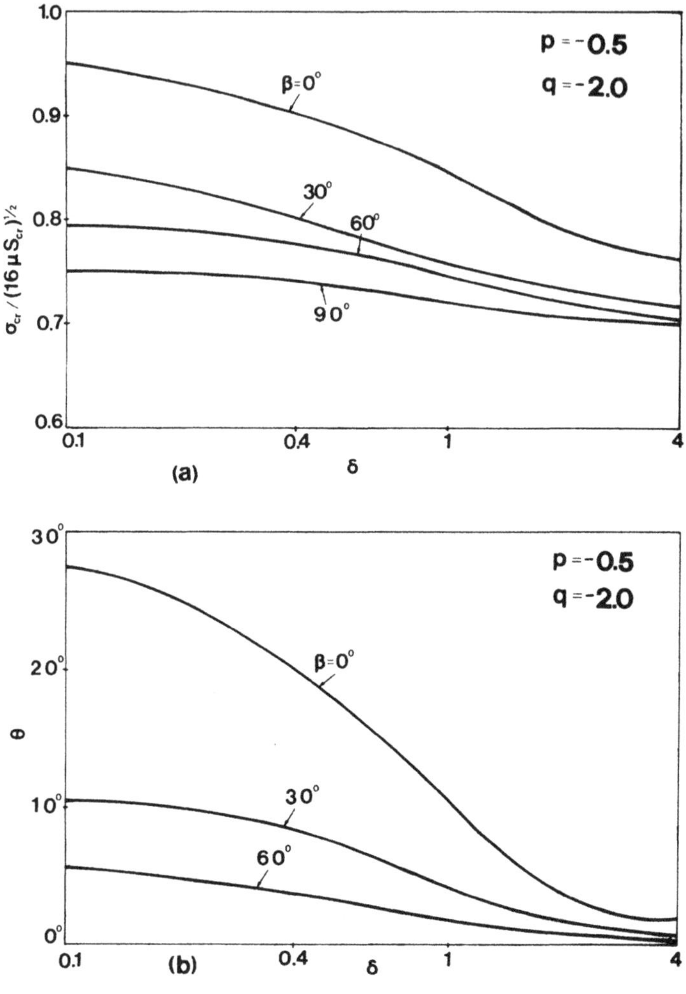

Figure 7.6. Variation of the quantity $\sigma_{cr}/(16\mu S_{cr})^{1/2}$ (a) and the crack extension angle θ (b) versus the distance δ for a cracked plate with a rigid boundary and $\kappa_1 = 2$. The crack is loaded by a uniform compressive stress along its surfaces and is inclined at an angle β equal to 90°, 60°, 30° and 0° with respect to the rigid boundary.

combinations of the bimaterial constants p and q: $p = 0.65$, $q = 0.80$ (Figure 7.7a), $p = 0.4$, $q = 0.6$ (Figure 7.7b), $p = -0.2$, $q = -0.6$ (Figure 7.8a) and $p = -0.35$, $q = -1.40$ (Figure 7.8b). Distance δ in Figures 7.6, 7.7 and 7.8 is set at 0.4. The crack extends from tip B in Figure 7.7, and from tip A in Figure 7.8. The values of the normalized critical stresses σ_{cr} are given in Figure 7.9 for an applied tensile load. The crack in this figure extends from tip A when p and q are positive, and from tip B when they are negative.

The above figures indicate clearly that the critical stress corresponding to the onset of crack extension, the specific crack tip from which extension initiates, and the crack extension angle are dependent on the crack location relative to the bimaterial interface

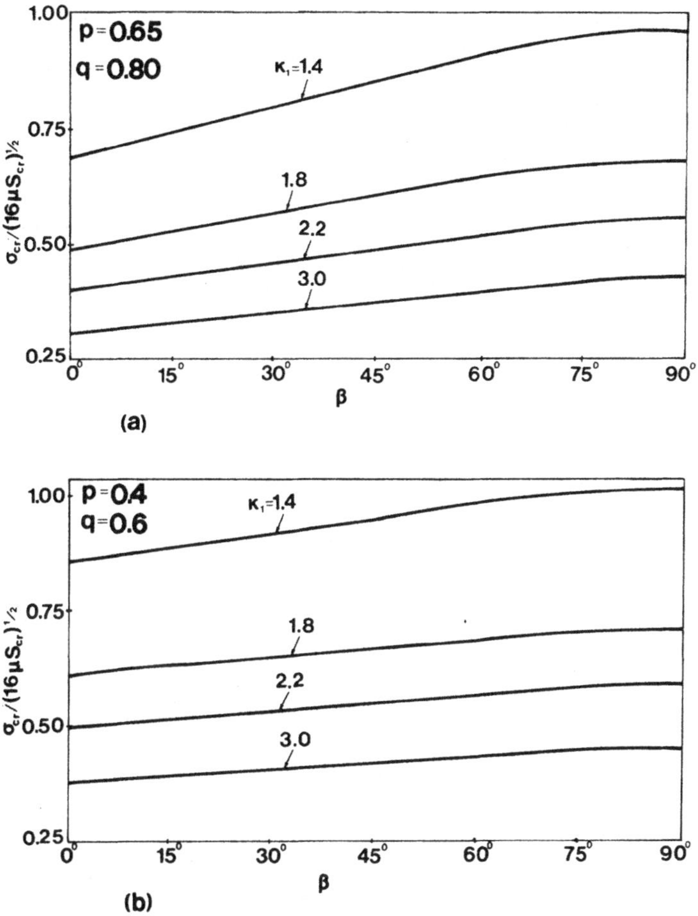

Figure 7.7. Variation of the quantity $\sigma_{cr}/(16\mu S_{cr})^{1/2}$ versus the crack inclination angle β with respect to the interface for a bimaterial plate with $p = 0.65$, $q = 0.80$ (a) and $p = 0.4$, $q = 0.6$ (b). The crack tip B is at a distance $\delta = 0.4$ from the interface and the crack is loaded by compressive stresses along its surfaces.

and on the material combinations of the composite plate. These results can be summarized as follows:

a. *The cracked plate with a stress-free boundary:*

(i) Under compressive applied stresses crack extension takes place from the tip nearest the boundary. Under tensile stresses the crack extends from the tip farthest from the boundary.

(ii) The fracture stress increases as the distance between the crack and the interface increases. This increase in stress is more abrupt under tensile than under compressive applied stresses.

(iii) The Poisson's ratio significantly affects the critical stress point for applied compressive stresses, while its influence is negligible for tensile stresses.

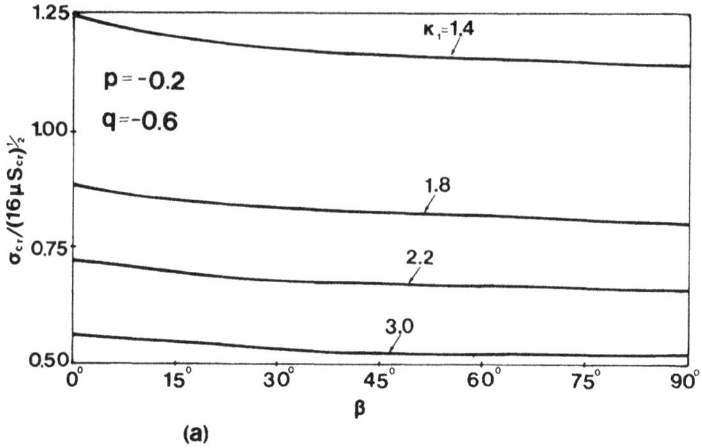

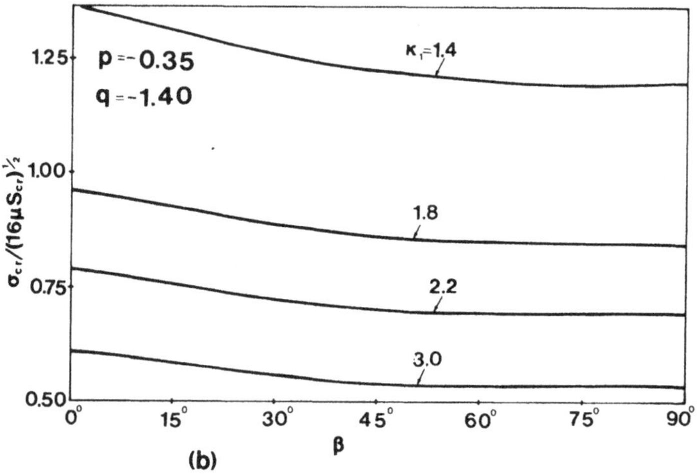

Figure 7.8. As in Figure 7.7 for $p = -0.2$, $q = -0.6$ (a) and $p = -0.35$, $q = -1.40$ (b).

(iv) For compressive stresses the critical stress decreases as the inclination of the crack with respect to the boundary decreases. Thus the critical stress reaches minimum for a crack parallel to the boundary. For tensile stresses no such rule can be stated.

(v) For compressive stresses the crack extension angle decreases and tends to zero as the distance between the crack and the boundary increases. Quite the reverse happens for tensile stresses.

(vi) For compressive stresses the crack extension angle increases as the crack inclination angle with respect to the boundary decreases. Quite the reverse happens for tensile stresses.

b. *The cracked plate with a rigid boundary (only compressive applied stresses are considered)*:

(i) Crack extension takes place from the tip farthest from the boundary.

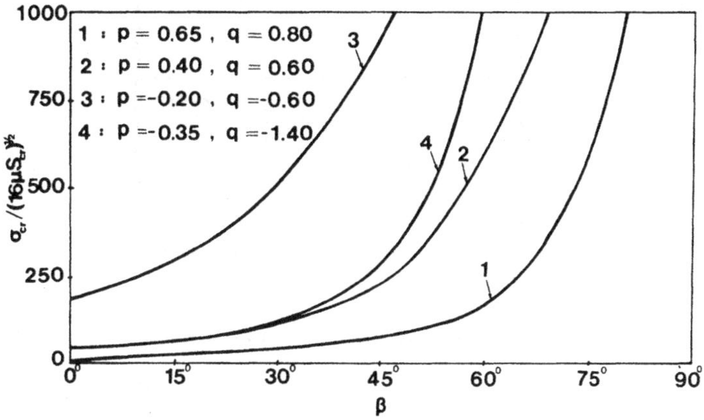

Figure 7.9. As in Figure 7.7 for tensile applied stresses and for the cases of Figures 7.7 and 7.8.

(ii) The critical stress increases as the distance between the crack and the boundary decreases.

(iii) The critical stress decreases as the crack inclination angle with respect to the boundary increases. Thus, the fracture stress is at a minimum when the crack is perpendicular to the interface.

(iv) The crack extension angle decreases and tends to zero as the distance between the crack and the boundary increases.

(v) The crack extension angle increases as the crack inclination angle with respect to the boundary decreases.

c. *The cracked composite plate with various bimaterial combinations:*

(i) The critical stress of fracture and the crack extension angle depend primarily on the elastic constants of the constituent materials of the composite plate.

(ii) At positive values of the bimaterial constants p and q the crack extends from the tip nearest the boundary for compressive and from the farthest tip, for tensile applied stresses. At negative values of the constants p and q, the relations are reversed.

(iii) For compressive applied stresses at positive values of the constants p and q, the fracture stress decreases as the crack inclination angle with respect to the interface decreases. At negative values of the constants p and q, the relations are again reversed.

(iv) For tensile applied stresses and for all values of the constants p and q the critical stress increases and tends to infinity as the crack inclination angle with respect to the interface increases.

The following general results apply to all the above cases:

(i) The critical stress and the crack extension angle increase as the Poisson's ratio ν of the plate increases.

(ii) The critical stress and the crack extension angle tend to the values of an isolated crack when the crack moves away from the bimaterial interface.

(iii) The critical stress for tensile applied stresses is many times greater than for compressive applied stresses.

The results of this section were taken from reference [37].

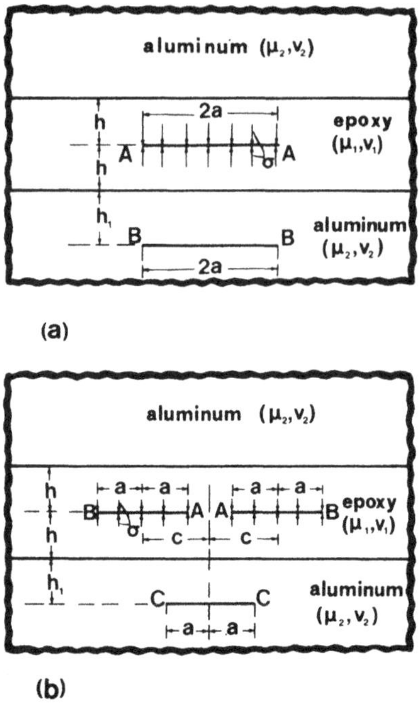

Figure 7.10. A three-layered composite made of two aluminum half-planes bonded through an epoxy layer. In case (a) one crack is embedded in the epoxy and one in the aluminum layer, while in case (b) the epoxy layer contains two cracks and the aluminum one.

7.3. A three-layered composite with cracks

The layered composite under study consists of two aluminum half-planes bonded by an epoxy layer (Figure 7.10). Two cases will be considered: in the first, one crack is embedded in the epoxy and another in the aluminum layer (Figure 7.10a); while in the second the epoxy layer contains two cracks and the aluminum one (Figure 7.10b). All cracks have the same length 2a and are parallel to the aluminum–epoxy interfaces. The elastic constants defined by the modulus of elasticity E and the Poisson's ratio ν take the values 10^7 psi and 0.3 for the aluminum, and 4.5×10^5 psi and 0.35 for the epoxy layer. The critical strain energy density factor S_{cr} is at $3.6\,\text{lb in}^{-1}$ for the aluminum [38] and at $5 \times 10^{-3}\,\text{lb in}^{-1}$ for the epoxy [35]. The composite plate is loaded by a uniform uniaxial compressive stress σ acting along the crack surfaces in the epoxy phase.

Ratwani and Gupta [19, 20] studied the problem of the layered composite, and gave the values of the stress intensity factors k_1 and k_2 for all tips of the existing cracks. Incorporating these values of k_1 and k_2 and the previously given values for the material constants μ, ν and S_{cr} into the equations of the strain energy density criterion (Section 2.2, page 11) the critical stresses for the onset of crack extension from the tips of all cracks are determined. It is found that fracture of the composite plate always takes place from the loaded crack in the epoxy layer.

Figure 7.11 presents the variation of the quantity $\sigma_{cr}(a/16\mu S_{cr})^{1/2}$ versus h_1/a for

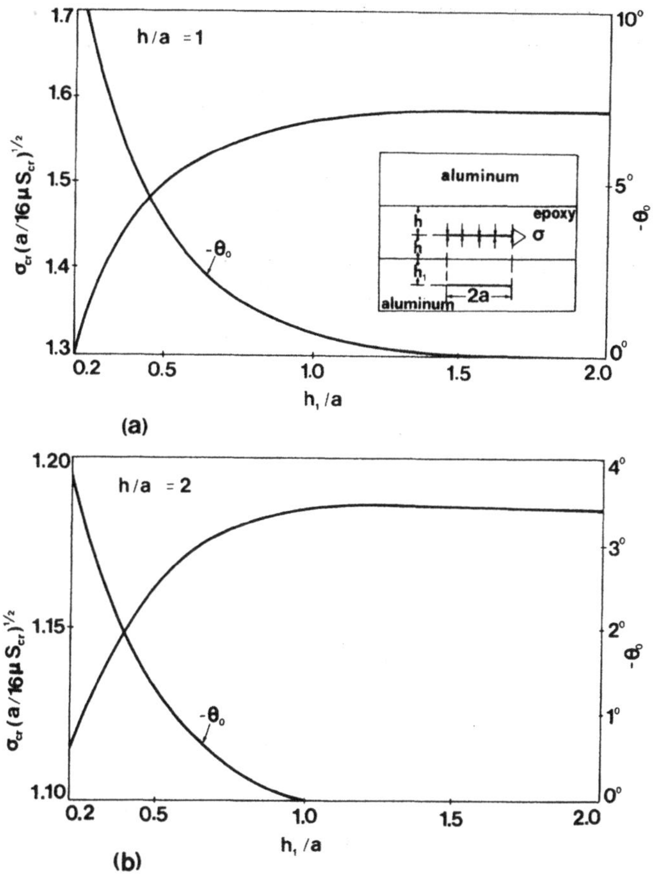

Figure 7.11. Variation of the quantity $\sigma_{cr}(a/16\mu S_{cr})^{1/2}$ and the crack extension angle $-\theta_0$ versus h_1/a for $h/a = 1$ (a) and $h/a = 2$ (b) for the case of Figure 7.10a. The loading of the plate consists of a uniform compressive stress along the surfaces of the crack in the epoxy layer. Fracture always takes place from the crack in the epoxy layer.

the layered composites illustrated in Figure 7.10a when $h/a = 1$ (Figure 7.11a) and $h/a = 2$ (Figure 7.11b). The constants μ and S_{cr} correspond to the epoxy layer, and conditions of plane strain prevail in the plate. In the same figures the values of the crack extension angle $-\theta_0$ are also presented. The values of the quantity $\sigma_{cr}(a/16\mu S_{cr})^{1/2}$ for Figure 7.10b for $h/a = 2$ and $h_1/a = 1$ are presented in Figure 7.12. The cracks AB in the epoxy layer extend from the outer tips B for $a/c < 0.85$ and from the inner tips for $a/c > 0.85$. The crack extension angle θ_0 for this case is approximately equal to zero.

Figures 7.11a and 7.11b support the following conclusions concerning the dependence of the critical stress σ_{cr} and the crack extension angle θ_0 on the geometrical configuration of the cracks illustrated by Figure 7.10a:

(i) The critical stress for failure of the composite plate decreases as the crack in the aluminum layer approaches the aluminum–epoxy interface and as the thickness of the

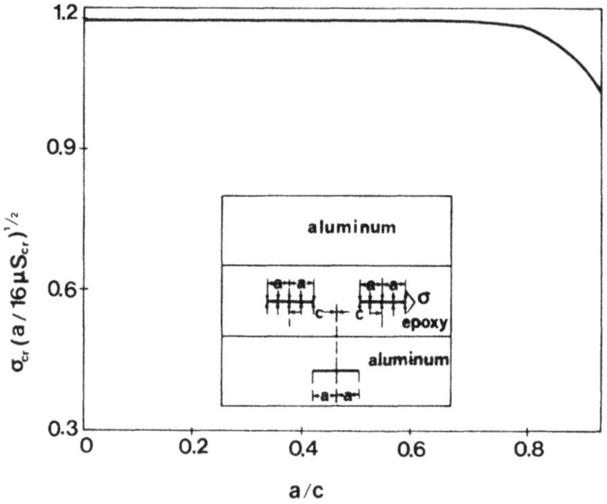

Figure 7.12. Variation of the quantity $\sigma_{cr}(a/16\mu S_{cr})^{1/2}$ versus a/c for the case of Figure 7.10b for $h/a = 2$ and $h_1/a = 1$. The loading of the plate consists of a uniform compressive stress along the surfaces of the cracks in the epoxy layer. Fracture takes place from the outer tips B of the cracks in the epoxy layer for $a/c < 0.85$ and from their inner tips for $a/c > 0.85$.

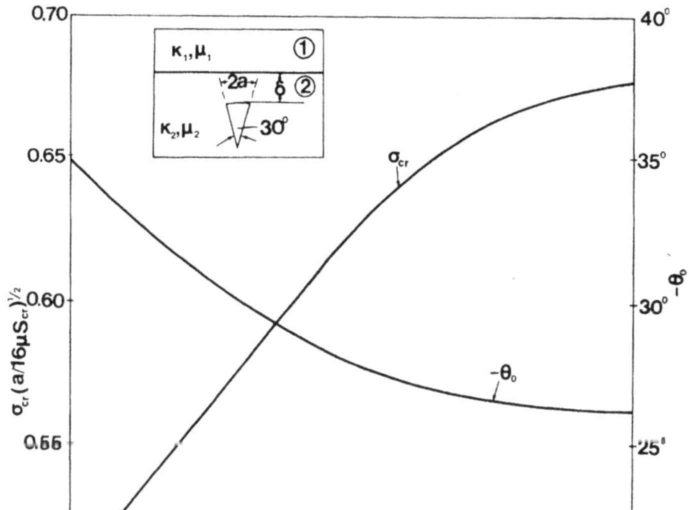

Figure 7.13. Variation of the quantity $\sigma_{cr}(a/16\mu S_{cr})^{1/2}$ and the crack extension angle $-\theta_0$ versus δ/a for a composite plate consisted of two different half-planes bonded together. A circular crack of length $2a$ exists in the one half-plane.

epoxy layer increases. This stress is always greater than the critical fracture stress for an epoxy layer of large dimensions.

(ii) The crack extension angle increases as the crack in the aluminum phase approaches the interface. It decreases as the thickness of the epoxy layer increases.

For more information on the results of fracture of the plate of Figure 7.10a the interested reader is referred to reference [39].

7.4. A bimaterial plate with a circular crack

Consider the bimaterial plate of Figure 7.13 made up of two bonded elastic half-planes containing a circular crack of length $2a$ and angle 30° in the second half-plane. Let δ be the distance from the crack to the interface of the half-planes. Assuming the elastic constants κ and μ at $\kappa_1 = 2.25$, $\kappa_2 = 2$ and $\mu_2 = 4\mu_1$, the normal σ_n and tangential σ_t stresses along the crack surfaces given by:

$$\sigma_n(s) = \sigma \cos^2 s \qquad \sigma_t(s) = \sigma \cos s \sin s \tag{7.2}$$

Ioakimidis and Theocaris [36] derived the values of the k_1 and k_2 stress intensity factors governing the stress field in the neighborhood of the crack tips. These values of k_1 and k_2 are used in the equations of the strain energy density criterion to obtain the critical stress σ_{cr} and the crack extension angle θ_0 (Figure 7.13). It can be observed from these results that the critical stress σ_{cr} increases with δ and tends to its limiting value for a circular crack lying in an infinite medium.

7.5. A bimaterial plate with a crack along the interface

Consider two homogeneous, isotropic, elastic materials which occupy the upper $(y > 0)$ and lower $(y < 0)$ half-planes with elastic constants $\mu_{1,2}$ and $\nu_{1,2}$ (Figure 7.14). The materials are bonded along the x-axis except for a finite segment of length $2a$ which forms an internal crack. The crack surfaces are free of loads and the composite plate is subjected to stresses at infinity.

Rice and Sih [2] and Erdogan [3] gave the following expressions for the singular stress components σ_{1x}, σ_{1y}, τ_{1xy} for the medium 1 $(0 < \theta < 180°)$ in the vicinity of the crack tip:

$$\begin{aligned}\sigma_{1x} = {} & \frac{K_I}{2\sqrt{2\pi r}} \left\{ \frac{1}{2} e^{-\beta(\pi-\theta)} \left[5 \cos\left(\frac{\theta}{2} + \beta \log\frac{r}{a}\right) + \cos\left(\frac{5\theta}{2} + \beta \log\frac{r}{a}\right) \right.\right. \\ & \left.\left. + 4\beta \sin\theta \cos\left(\frac{3\theta}{2} + \beta \log\frac{r}{a}\right)\right] - e^{\beta(\pi-\theta)} \cos\left(\frac{\theta}{2} - \beta \log\frac{r}{a}\right)\right\} \\ & - \frac{K_{II}}{2\sqrt{2\pi r}} \left\{ \frac{1}{2} e^{-\beta(\pi-\theta)} \left[5 \sin\left(\frac{\theta}{2} + \beta \log\frac{r}{a}\right) + \sin\left(\frac{5\theta}{2} + \beta \log\frac{r}{a}\right) \right.\right. \\ & \left.\left. + 4\beta \sin\theta \sin\left(\frac{3\theta}{2} + \beta \log\frac{r}{a}\right)\right] + e^{\beta(\pi-\theta)} \sin\left(\frac{\theta}{2} - \beta \log\frac{r}{a}\right)\right\}\end{aligned} \tag{7.3a}$$

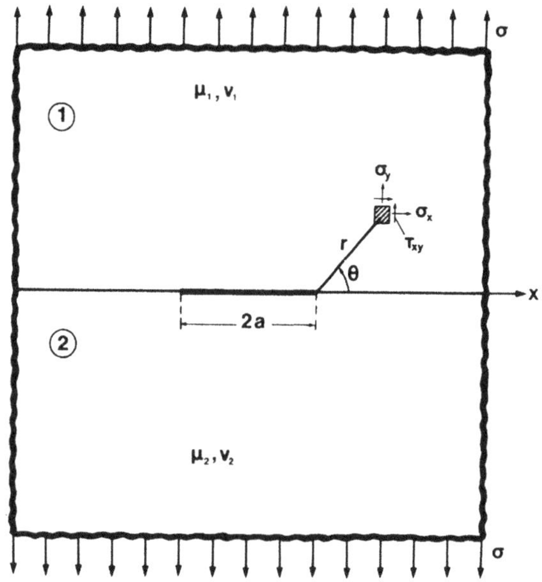

Figure 7.14. A bimaterial plate with a crack along the interface.

$$\sigma_{1y} = \frac{K_I}{2\sqrt{2\pi r}}\left\{\frac{1}{2}e^{-\beta(\pi-\theta)}\left[3\cos\left(\frac{\theta}{2}+\beta\log\frac{r}{a}\right)-\cos\left(\frac{5\theta}{2}+\beta\log\frac{r}{a}\right)\right.\right.$$
$$\left.\left.-4\beta\sin\theta\cos\left(\frac{3\theta}{2}+\beta\log\frac{r}{a}\right)\right]+e^{\beta(\pi-\theta)}\cos\left(\frac{\theta}{2}-\beta\log\frac{r}{a}\right)\right\}$$
$$+\frac{K_{II}}{2\sqrt{2\pi r}}\left\{\frac{1}{2}e^{-\beta(\pi-\theta)}\left[-3\sin\left(\frac{\theta}{2}+\beta\log\frac{r}{a}\right)+\sin\left(\frac{5\theta}{2}+\beta\log\frac{r}{a}\right)\right.\right.$$
$$\left.\left.+4\beta\sin\theta\sin\left(\frac{3\theta}{2}+\beta\log\frac{r}{a}\right)\right]+e^{\beta(\pi-\theta)}\sin\left(\frac{\theta}{2}-\beta\log\frac{r}{a}\right)\right\} \quad (7.3b)$$

$$\tau_{1xy} = \frac{K_I}{2\sqrt{2\pi r}}\left\{\frac{1}{2}e^{-\beta(\pi-\theta)}\left[\sin\left(\frac{\theta}{2}+\beta\log\frac{r}{a}\right)+\sin\left(\frac{5\theta}{2}+\beta\log\frac{r}{a}\right)\right.\right.$$
$$\left.\left.+4\beta\sin\theta\sin\left(\frac{3\theta}{2}+\beta\log\frac{r}{a}\right)\right]-e^{\beta(\pi-\theta)}\sin\left(\frac{\theta}{2}-\beta\log\frac{r}{a}\right)\right\}$$
$$+\frac{K_{II}}{2\sqrt{2\pi r}}\left\{\frac{1}{2}e^{-\beta(\pi-\theta)}\left[\cos\left(\frac{\theta}{2}+\beta\log\frac{r}{a}\right)+\cos\left(\frac{5\theta}{2}+\beta\log\frac{r}{a}\right)\right.\right.$$
$$\left.\left.+4\beta\sin\theta\cos\left(\frac{3\theta}{2}+\beta\log\frac{r}{a}\right)\right]+e^{\beta(\pi-\theta)}\cos\left(\frac{\theta}{2}-\beta\log\frac{r}{a}\right)\right\}. \quad (7.3c)$$

In the above relations r, θ are the polar coordinates of the point considered, β is a bimaterial constants given by:

$$\beta = \frac{1}{2\pi}\log\beta_0 \quad (7.4a)$$

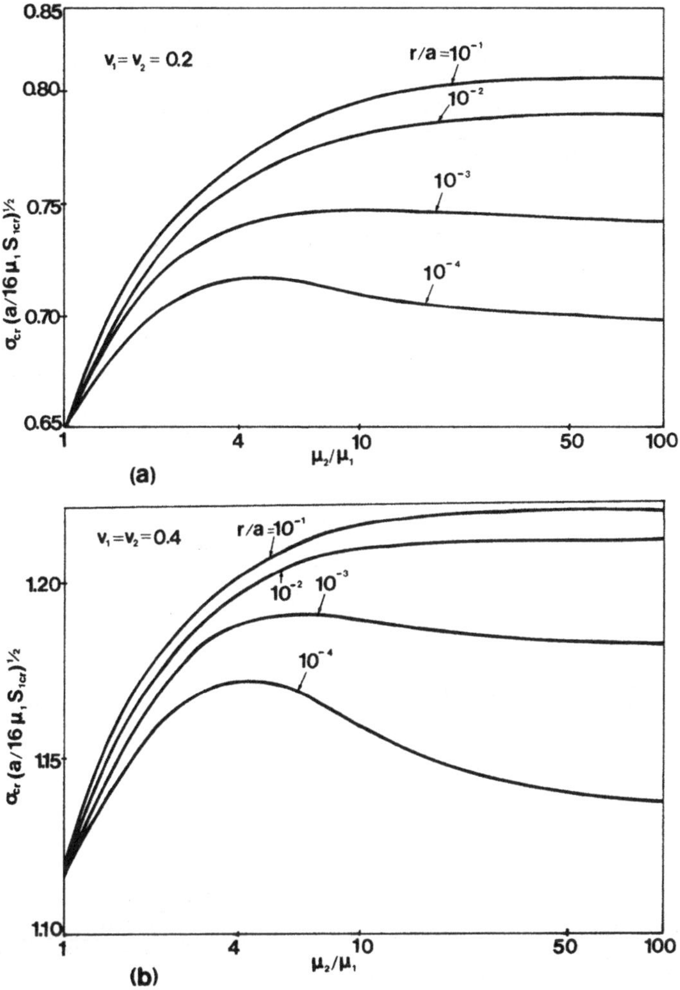

Figure 7.15. Variation of the normalized critical stress of fracture $\sigma_{cr}(a/16\mu_1 S_{1cr})^{1/2}$ versus μ_2/μ_1 for $\nu_1 = \nu_2 = 0.2$ (a) and $\nu_1 = \nu_2 = 0.4$ (b). The plate is subjected to a tensile stress and the radius r of the core region takes the values $r/a = 10^{-4}$, 10^{-3}, 10^{-2} and 10^{-1}. The crack always extends into material 1.

with

$$\beta_0 = \frac{\mu_1 + \kappa_1\mu_2}{\mu_2 + \kappa_2\mu_1}. \tag{7.4b}$$

The K_1 and K_2 stress intensity factors are independent of the coordinates r, θ, but depend on the elastic constants of the two media and the loading conditions of the plate.

From the above definition of the bimaterial constant β_0 it is established that $1/3 \leqslant \beta_0 \leqslant 3$. The values $\beta_0 = 1/3$ and 3 correspond to $\mu_2/\mu_1 = 0$ and ∞ respectively with

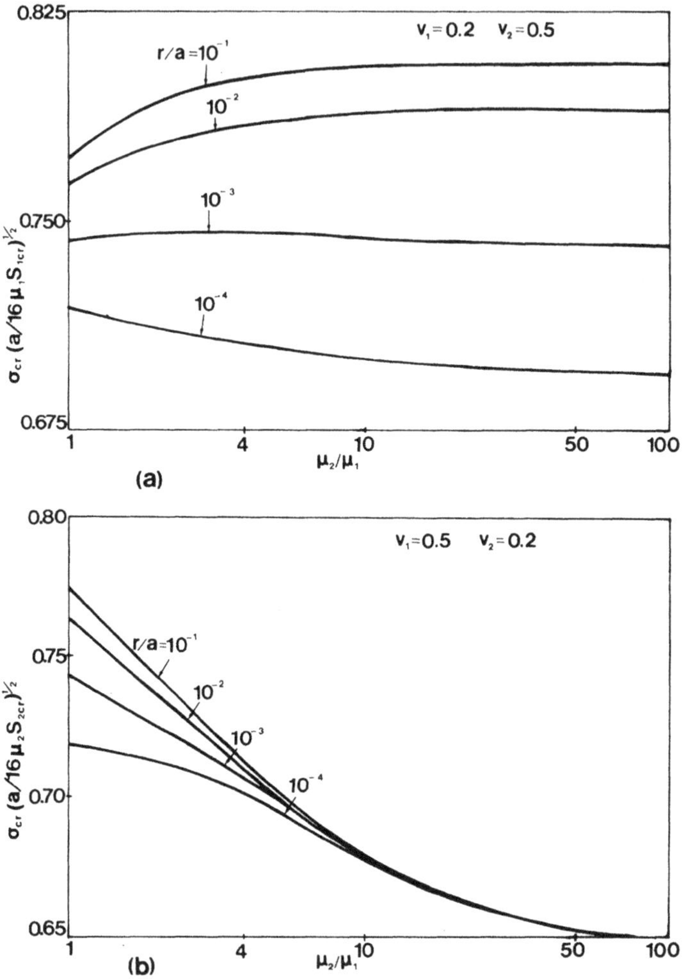

Figure 7.16. Variation of the normalized critical stress of fracture $\sigma_{cr}(a/16\mu_{1,2}S_{1,2cr})^{1/2}$ versus μ_2/μ_1 for $\nu_1 = 0.2$, $\nu_2 = 0.5$ (a) and $\nu_1 = 0.5$, $\nu_2 = 0.2$ (b). The plate is subjected to a tensile stress and the radius r of the core region takes the values 10^{-4}, 10^{-3}, 10^{-2} and $10^{-1}a$. The crack extends into material 1 for case a and into material 2 for case b.

$\kappa_{1,2} = 3$ (corresponding to the value of the Poisson's ratio $\nu = 0$ under plane strain conditions). As it is shown from relation (7.4b) when the materials 1 and 2 are interchanged the constant β_0 takes the value $1/\beta_0$.

When both materials have the same elastic properties the bimaterial constant β is equal to zero and relations (7.3) coincide with relations (2.1) for a crack in an infinite plate.

When the plate is subjected to a uniform uniaxial stress σ along the y-axis at large distances from the crack the stress intensity factors K_1 and K_2 are given by:

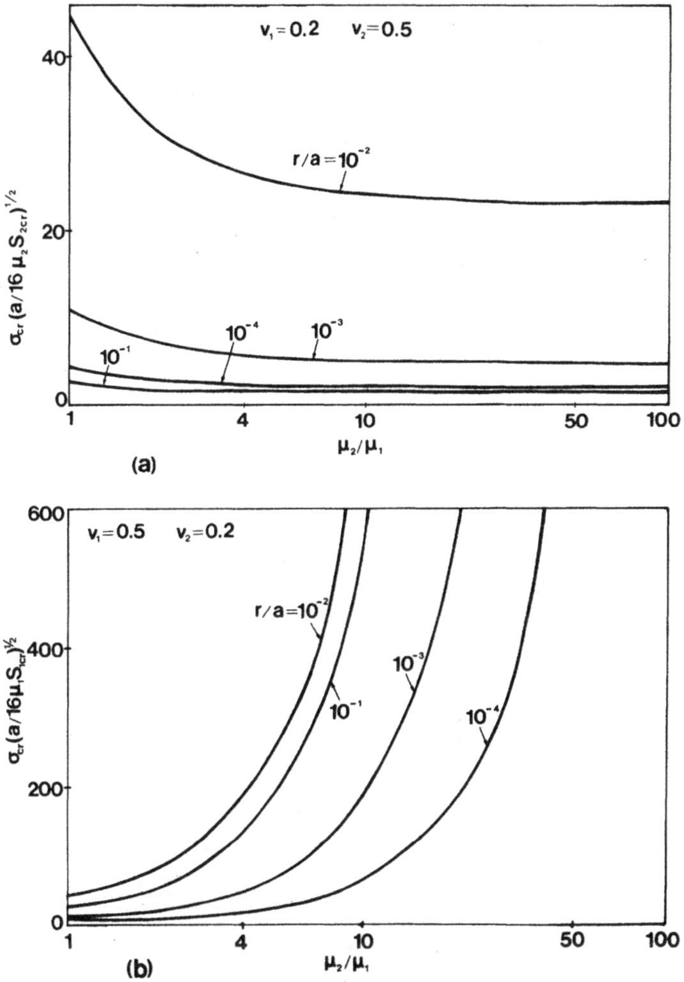

Figure 7.17. As in Figure 7.16 for compressive applied loads. The crack extends into material 2 for case (a) and into material 1 for case (b).

$$K_1 = \frac{\cos(\beta \log 2a) + 2\beta \sin(\beta \log 2a)}{\cosh \pi\beta} \sigma\sqrt{\pi a} \tag{7.5a}$$

$$K_2 = -\frac{\sin(\beta \log 2a) - 2\beta \cos(\beta \log 2a)}{\cosh \pi\beta} \sigma\sqrt{\pi a}. \tag{7.5b}$$

The brittle fracture characteristic quantities of the plate subjected to the stress σ are determined by introducing the values of the stress components from relations (7.3) and K_1 and K_2 from (7.5) into the general equations (1.2) and (1.4) of the strain energy density criterion. The value of the radius r of the core region surrounding the crack tip must first be found in order to determine the strain energy density factor S.

As proposed by Sih [40], r can be approximately determined by assuming that

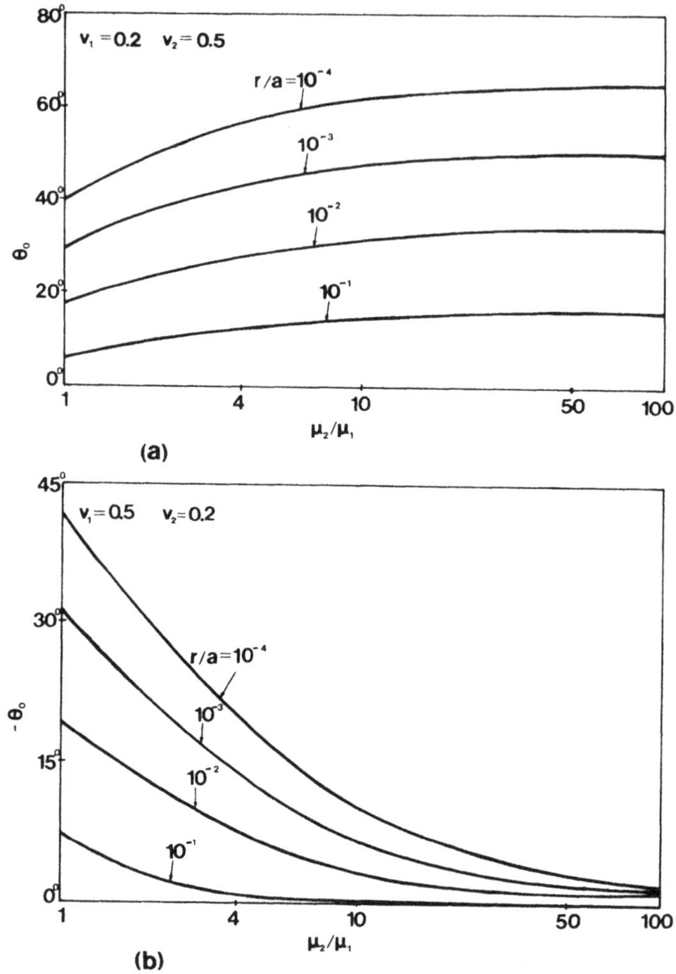

Figure 7.18. Variation of the crack extension angle θ_0 versus μ_2/μ_1 for $\nu_1 = 0.2$, $\nu_2 = 0.5$ (a) and $\nu_1 = 0.5$, $\nu_2 = 0.2$ (b). The plate is subjected to a tensile stress and the radius r of the core region takes the values $r/a = 10^{-4}$, 10^{-3}, 10^{-2} and 10^{-1}.

incipient fracture coincides with the critical value of the strain energy density function $\mathrm{d}W/\mathrm{d}V$ in an element at a distance r from the crack tip, i.e. $\mathrm{d}W/\mathrm{d}V = C$. The value of C can be determined from a simple tension test. For a linear material C has the value $\sigma_u^2/2E$, where σ_u is the ultimate strength and E the Young's modulus. It is thus found that for most engineering metal alloys r ranges from 10^{-3} to 10^{-2}in. Outside the core region the stress field is described by the analytical solution, while inside the core region such a description is impossible. Thus, the boundary of the core region separates the outside elastic material from the inside material without specified mechanical properties. The size of the core region reflects a basic material property at the microscopic scale.

The radius r of the core region was obtained equal to 10^{-4}, 10^{-3}, 10^{-2} and $10^{-1}a$.

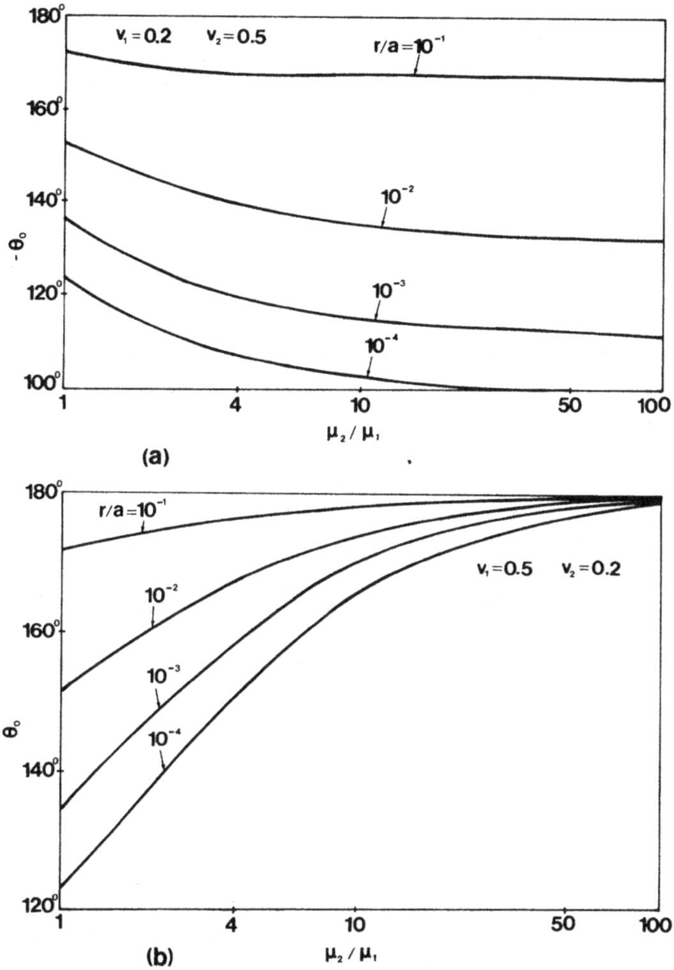

Figure 7.19. As in Figure 7.18 for compressive applied loads.

These values of r are generally outside the critical region surrounding the crack tip in which the stresses present the well-known oscillatory behavior. When the minimum values of S coincide with their critical values S_{1cr} and S_{2cr} for materials 1 and 2 the critical values σ_{cr} of the applied stress σ for crack extension can be determined.

Figure 7.15a presents the variation of the quantity $\sigma_{cr}(a/16\mu_1 S_{1cr})^{1/2}$ versus μ_2/μ_1 for $\nu_1 = \nu_2 = 0.2$ and $r/a = 10^{-4}$, 10^{-3}, 10^{-2} and 10^{-1} when the stress σ is tensile. σ_{cr} is the critical value of σ for crack extension and S_{1cr} is the critical value of S for material 1. It was found that under these conditions the crack always extends into material 1. The values of the same quantity for $\nu_1 = \nu_2 = 0.4$, $\nu_1 = 0.2$, $\nu_2 = 0.5$ and $\nu_1 = 0.5$, $\nu_2 = 0.2$ under tensile applied loads are presented in Figures 7.15b, 7.16a and 7.16b respectively. It was found that the crack extends into material 1 for Figures 7.15b and 7.16a and into material 2 for Figure 7.16b. When the Poisson's ratios ν_1 and ν_2 are

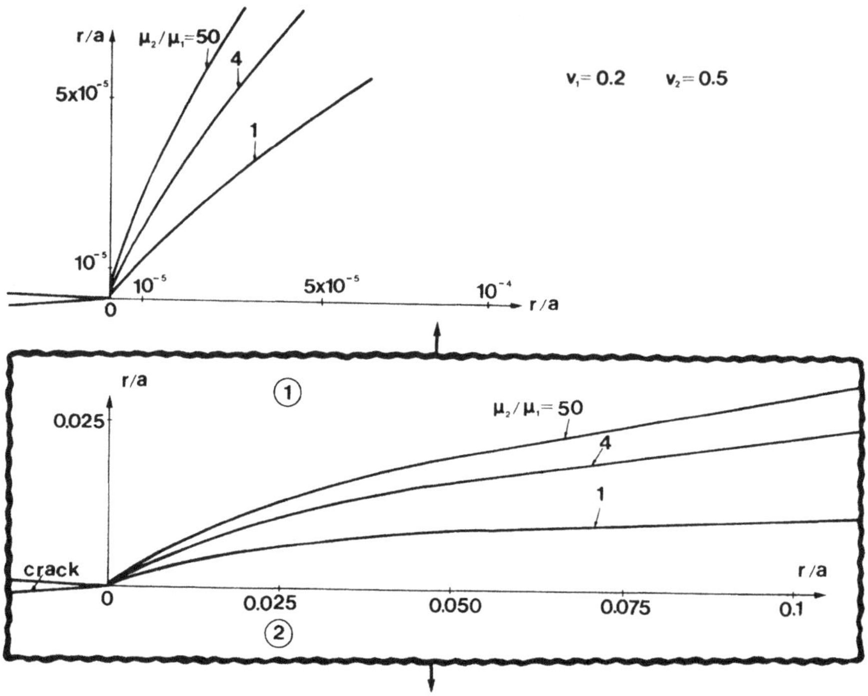

Figure 7.20. Fracture trajectories for a bimaterial plate subjected to tensile loads with $\nu_1 = 0.2$, $\nu_2 = 0.5$ and $\mu_2/\mu_1 = 1$, 4 and 50. The detailed form of the trajectories in the close neighborhood of the crack tip is shown in the upper left part of the figure.

equal the values of the quantity $\sigma_{cr}(a/16\mu_1 S_{1cr})^{1/2}$ for $\mu_1 = \mu_2$ are 0.645 and 1.118, corresponding to a crack in an infinite plate with $\nu = 0.2$ and 0.4 respectively. From Figures 7.15 and 7.16 the trend of the increased allowable loading with the increased radius of the core region r is established.

Figure 7.17 illustrates the values of the quantity $\sigma_{cr}(a/16\mu S_{cr})^{1/2}$ for $\nu_1 = 0.2$, $\nu_2 = 0.5$ and $\nu_1 = 0.5$, $\nu_2 = 0.2$ under compressive loads. The crack extends into material 2 for Figure 7.17a and into material 1 for Figure 7.17b. By comparing the values of the critical stress for tensile and compressive applied loads, it is concluded that this stress is always larger for compressive than for tensile loads. This result is also found when a crack in a single plate is subjected to a uniform stress perpendicular to the crack plane (Section 2.3d, page 18). When compressive loads are applied, note that caution must be exercised in avoiding interpenetration of notch faces.

The values of the crack extension angle θ_0 in Figures 7.16 and 7.17 are presented in Figures 7.18 and 7.19. Based on these figures the fracture trajectories for tensile and compressive loads can be determined. Indeed, for brittle fracture behavior with the fracture growth sudden and unstable it does not seem unreasonable to assume that the fracture path can be determined from the conditions existing prior to fracture initiation. Since the material does not have time to redistribute the stresses during the fracture process, the fracture path can be predetermined with high accuracy. A sound

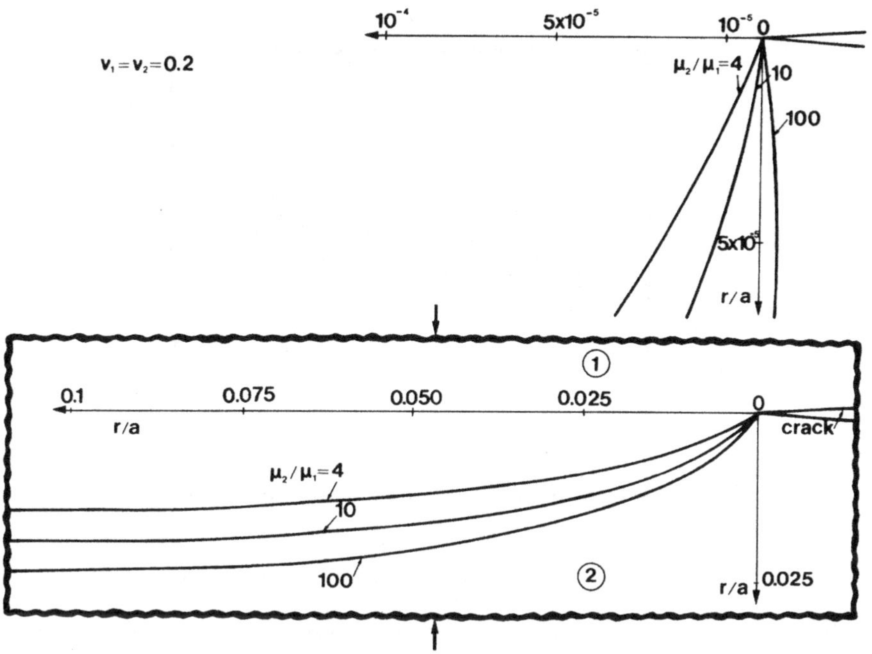

Figure 7.21. As in Figure 7.20 for compressive applied loads and $\nu_1 = \nu_2 = 0.2$.

justification in favor of such behavior comes from the case of brittle fracture of blunt notches or line cracks at low angles of loading where the loads required to initiate fracture are relatively high. In such cases there exists a high density energy field near and far from the initial crack extension area which is responsible for rapid fracture when other physical resources exist to inhibit fracture from propagation.

Figure 7.20 presents the fracture trajectories for $\nu_1 = 0.2$, $\nu_2 = 0.5$ and $\mu_2/\mu_1 = 1$, 4 and 50 in a bimaterial plate subjected to tensile loads. In the upper left side of this figure a detailed sketch of fracture trajectories close to the crack tip is shown. It is observed that as distance from the crack tip increases the crack tends to become parallel to its initial direction. This result is in accordance with the findings of reference [15].

Similarly, Figure 7.21 presents the fracture trajectories for compressive applied loads for $\nu_1 = \nu_2 = 0.2$ and $\mu_2/\mu_1 = 4$, 10 and 100. We observe that the crack propagates in a direction opposite that of a crack propagating under tensile loads. However, as in the case of tension the crack tends to become parallel to its initial direction as the distance from the crack tip increases.

We can thus conclude from the above developments that the fracture characteristic quantities are dependent on the material properties of the two phases of the bimaterial plate in the following ways:

(i) For tensile loads when $\nu_1 \leqslant \nu_2$ the crack extends into material 1, while when for $\nu_1 > \nu_2$ it extends into material 2. Quite the reverse happens for compressive loads.

(ii) The critical stress for crack extension increases as the radius r of the core region surrounding the crack tip increases. Determination of r, a material constant, enables the calculation of the critical stress.

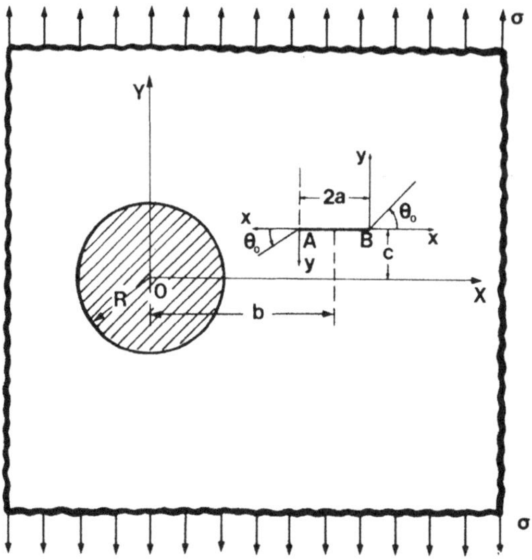

Figure 7.22. Geometry of the cracked plate with a hole or an inclusion.

(iii) The critical stress for crack extension is always larger for compressive than for tensile applied loads.

(iv) The crack extension angle θ_0 for tensile loads decreases and tends to zero as the radial distance r/a from the crack tip increases. For compressive loads the angle θ_0 tends to $\pm 180°$ as r/a increases.

(v) The fracture path for both tensile and compressive loads tends to become parallel to the crack axis as r/a increases. However, the crack extends in opposite directions for tensile and compressive loads.

7.6. Interaction between a crack and a circular inclusion

Consider the case of an elastic plate with a circular inclusion of radius R, containing a crack of length $2a$ equal to R (Figure 7.22). The plate is subjected to a uniaxial uniform stress σ perpendicular to the crack axis. We will consider below the situations when the inclusion degenerates to a hole, and when it is extremely stiff with respect to the matrix (a metallic inclusion embedded into an epoxy matrix).

a. *Interaction between a crack and a hole:* This problem has been considered by Erdogan *et al.* [33] who gave the values of the opening-mode, k_1, and sliding-mode, k_2, stress intensity factors for both crack tips A and B. Using these values of k_1 and k_2 in combination with the strain energy density criterion the critical values of the applied stress σ_{cr}^A, σ_{cr}^B for possible extension of the crack AB from either of its tips A or B are determined. The fracture stress of the plate is equal to the minimum value of σ_{cr}^A and σ_{cr}^B.

Figure 7.23 presents the variation of the normalized critical stress $\sigma_{cr}(a/16\mu S_{cr})^{1/2}$ versus b/a for the symmetrical case when the crack moves on the diameter OX of the hole perpendicular to the applied stress. κ takes the values: 1.4, 1.8, 2.2 and 3.0. Due

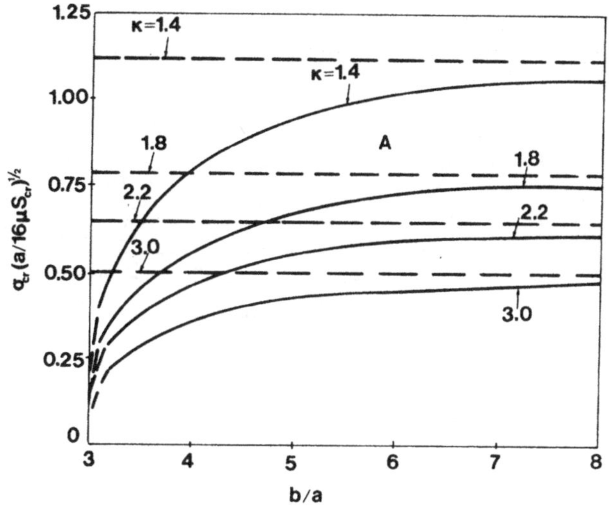

Figure 7.23. Variation of the quantity $\sigma_{cr}(a/16\mu S_{cr})^{1/2}$ versus b/a for the case of a cracked plate perforated by a hole with $c = 0$ for $\kappa = 1.4, 1.8, 2.2$ and 3.0. The values of the above quantity for an isolated crack are indicated by dotted lines. The crack always extends from its tip A.

to the symmetry of the problem only k_1 exists for both crack tips and therefore the crack extends in its own plane. It was found that the crack always extends from tip A located close to the hole. In order to obtain an estimate of the effect of the hole on the value of the fracture stress of the cracked plate the case of an isolated crack of length $2a$ is also considered. For the latter, the values of the quantity $\sigma_{cr}(a/16\mu S_{cr})^{1/2}$ are determined and plotted in Figure 7.23 by dotted lines. It can be observed that the fracture stress for the perforated plate is always smaller than the corresponding stress for the plate without the hole, but that both stresses approach each other as the crack moves away from the hole. Thus, the presence of the hole greatly diminishes the fracture stress of the cracked plate when the crack is in close proximity to the hole, while as their distance increases the influence of the hole becomes insignificant.

The variation of the normalized critical stress $\sigma_{cr}(a/16\mu S_{cr})^{1/2}$ and the fracture angle θ_0 versus b/a for $c = 2.2a$ is presented in Figure 7.24. Crack extension for $b/a < 3$ takes place from tip B, located at a greater distance from the hole, while for $b/a > 3$ it extends from tip A. Dotted lines separate the area in which crack extension takes place from tip A from that in which the crack extends from tip B. In Figure 7.24a dotted lines also represent the values of the critical stress for a plate with a single crack. We observe that for small values of b/a the fracture stress of the plate with the hole is greater than the fracture stress of the plate without the hole, while for larger values of b/a this rule is reversed. Thus, for small values of b/a the existence of the hole increases the fracture stress of the cracked plate. This apparent paradox can be explained by the fact that when an adequate overlap between the hole and the crack exists, the energy stored in the plate by the external loads is shared between the hole and the crack, and therefore the intensity of the stress field around the crack for the perforated plate is smaller than the corresponding intensity for the plate without the hole.

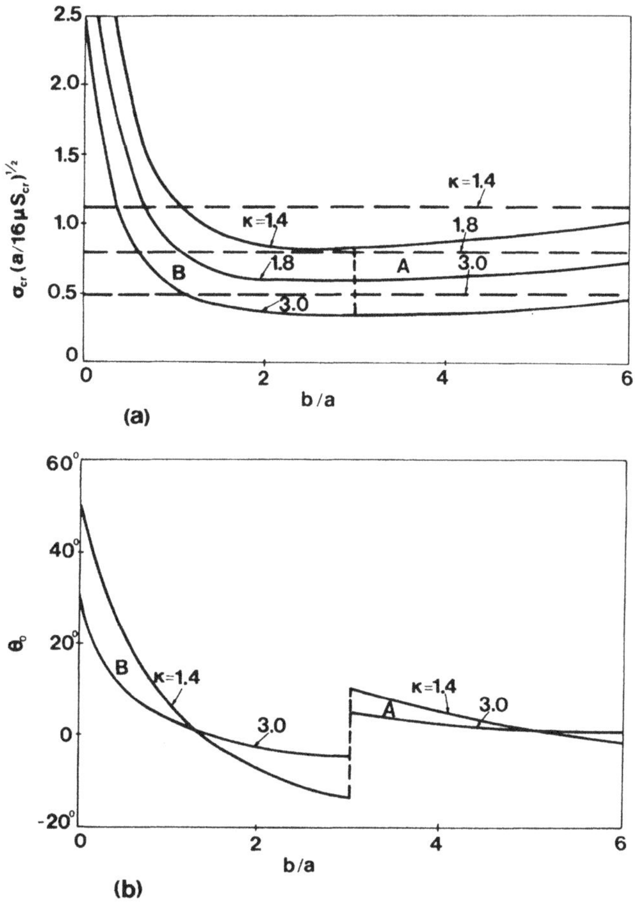

Figure 7.24. As in the case of Figure 7.23 with $c = 2.2a$. Regions in which the crack extends from its tip A or B are separated by dotted lines.

While in Figures 7.23 and 7.24 the crack moves on lines perpendicular to the applied stress, Figure 7.25 presents the variation of the quantities σ_{cr} and θ_0 when the crack moves in the direction of the applied stress for $b = 3a$ (the left crack tip touches the rim of the hole for $c = 0$). Again, dotted lines separate the area in which crack extension takes place from the tip A from the area in which it starts from tip B. Furthermore, the values of the critical stress for a single crack in a plate were drawn by dotted lines in Figure 7.25(a).

The above analysis leads to the following general conclusions regarding the dependence of the fracture stress and the crack extension angle on the relative location of the crack and the hole and the material properties of the plate:

(i) Determination of the tip from which the crack extends depends on the relative location of the crack and the hole. Thus, for movement of the crack on horizontal lines close to the diameter of the hole with small values of b/a the crack extends from the tip A toward the hole. As the distance b/a increases the crack propagates away

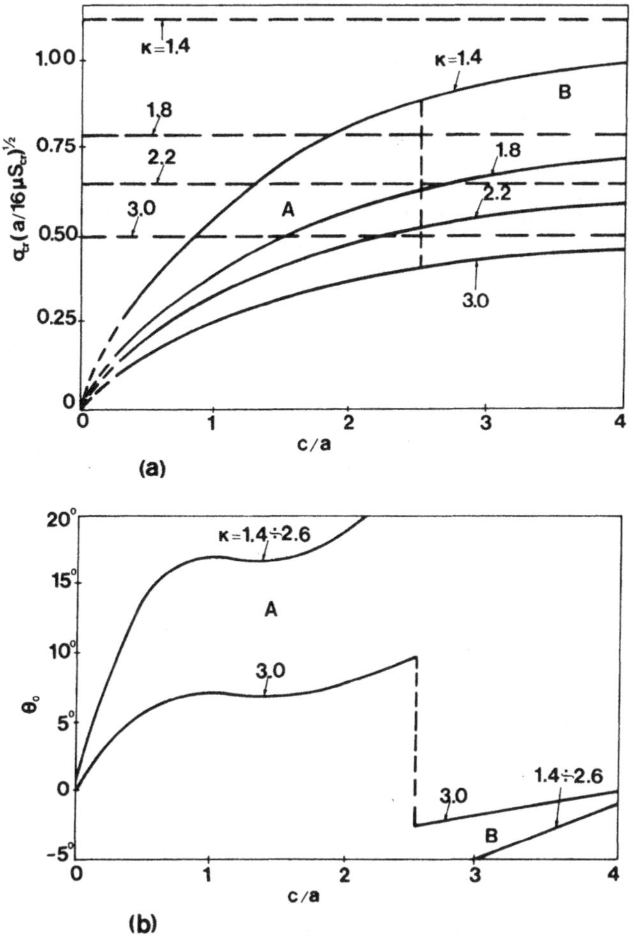

Figure 7.25. Variation of the quantity $\sigma_{cr}(a/16\mu S_{cr})^{1/2}$ (a) and the crack extension angle θ_0 (b) versus c/a for the case of a cracked plate perforated by a hole with $b = 3a$. The values of the quantity $\sigma_{cr}(a/16\mu S_{cr})^{1/2}$ for an isolated crack are also indicated. Regions in which the crack extends from its tip A or B are separated by dotted lines.

from the hole. However, as the distance c/a increases the crack extends from its tip B for small values of b/a, while for greater values of b/a the crack extension takes place from tip A in the direction of the hole. When the crack moves in the direction of the applied stress and it is completely overlapped by the hole it always propagates toward the hole; however, when the crack and the hole do not overlap, the value of c/a determines whether the crack extends from the tip A or B, and toward or away from the hole.

(ii) The crack extension angle is larger when the crack is close to the hole, while as the crack recedes from the hole and moves to infinity, this angle decreases and tends to zero.

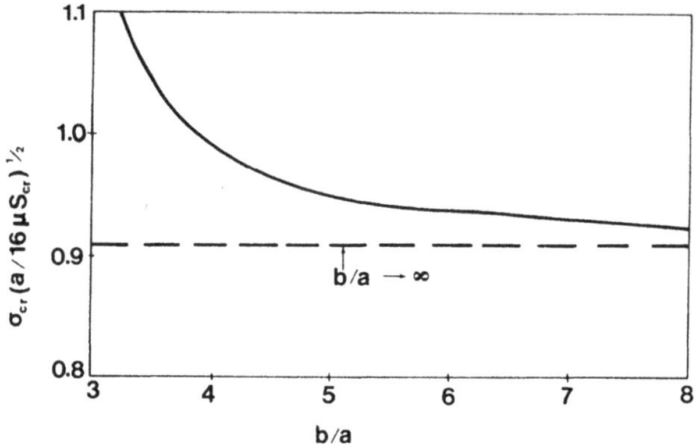

Figure 7.26. Variation of the quantity $\sigma_{cr}(a/16\mu S_{cr})^{1/2}$ versus b/a for the case of a cracked plate with an inclusion for $c = 0$. It is: $\mu_2/\mu_1 = 23$, $\kappa_1 = 1.6$, $\kappa_2 = 1.8$. The value of the above quantity for an isolated crack is also indicated.

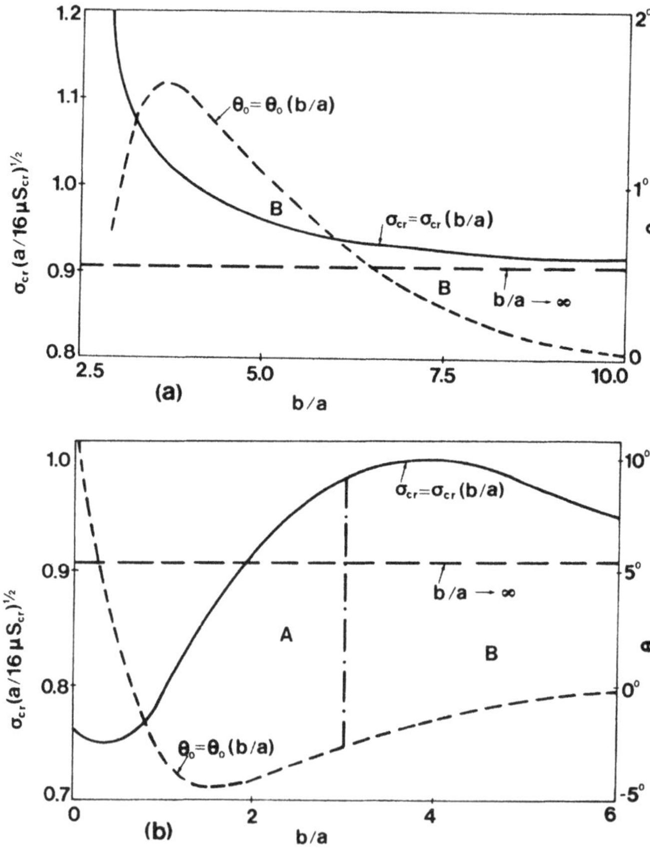

Figure 7.27. Variation of the quantity $\sigma_{cr}(a/16\mu S_{cr})^{1/2}$ (continuous lines) and the crack extension angle θ_0 (dotted lines) versus b/a for the case of a cracked plate with an inclusion for $c = a$. The value of the quantity $\sigma_{cr}(a/16\mu S_{cr})^{1/2}$ for an isolated crack is also indicated (a). As in case (a) for $c = 2.2a$ (b). The material constants take the same values as in the case of Figure 7.26.

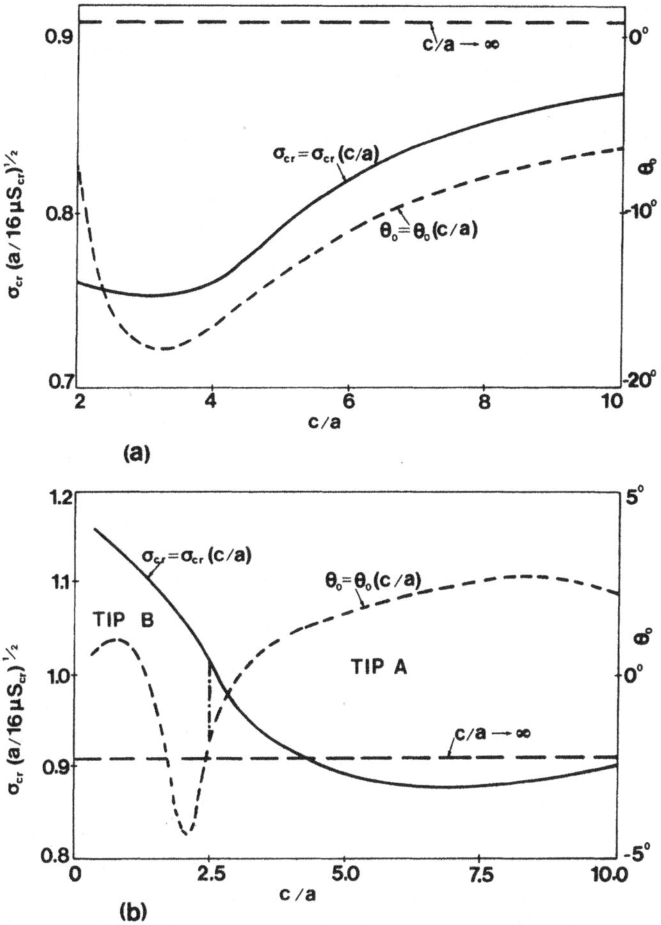

Figure 7.28. Variation of the quantity $\sigma_{cr}(a/16\mu S_{cr})^{1/2}$ (continuous lines) and the crack extension angle θ_0 (dotted lines) versus c/a for the case of a cracked plate with an inclusion for $b = 0$. The value of the quantity $\sigma_{cr}(a/16\mu S_{cr})^{1/2}$ for an isolated crack is also indicated (a). As in case (a) for $b = 3a$ (b). The material constants take the same values as in the case of Figure 7.26.

(iii) The existence of the hole greatly reduces the fracture strength of the plate when the crack and the hole do not overlap. The fracture strength is very low when the crack is close to the hole, and it increases as the distance between the crack and the hole increases. On the other hand, when overlapping exists the presence of the hole increases the fracture strength of the plate. This increase corresponds to the size of the overlapping lengths. In all cases as the crack recedes from the hole the fracture strength tends to that of a plate with an isolated crack.

b. *Interaction between a crack and an inclusion:* Consider a metallic inclusion embedded into an epoxy matrix, with values of $\mu_2/\mu_1 = 23$, $\kappa_1 = 1.6$ and $\kappa_2 = 1.8$, (subscript 1 refers to the matrix and subscript 2, to the inclusion). Proceeding as in Section 7.6a, the critical stress and the crack extension angle are determined.

Figure 7.26 presents the variation of the normalized critical fracture stress σ_{cr}

versus b/a for the symmetrical situation when the crack moves on the horizontal line passing through the center of the inclusion. Due to the symmetry of the problem the crack always extends along its own plane. We observe that the fracture stress σ_{cr} decreases as the distance between the crack and the inclusion increases, and that σ_{cr} is always greater than the fracture stress of the plate without the inclusion (shown in figure by dotted lines). Therefore, contrary to the case of the hole, the presence of the inclusion increases the fracture strength of the cracked plate.

The variation of the quantities $\sigma_{cr}(a/16\mu S_{cr})^{1/2}$ and θ_0 versus b/a is presented in Figure 7.27a for $c = a$ and in Figure 7.27b for $c = 2.2a$ with continuous and dotted lines respectively, while the same quantities for $b = 0$ and $b = 3a$ are presented in Figures 2.28a and 2.28b. These figures illustrate the dependence of the fracture stress σ_{cr} and the crack extension angle θ_0 on the geometry of the plate.

As in the previous case of the interaction between a crack and a hole the following conclusions can be derived:

(i) When the cracked composite material contains an inclusion the crack always extends from the tip other than the one from which it extends when the material contains a hole. Therefore, all laws governing the dependence of the more vulnerable tip from which the crack propagates on the geometrical situation of the plate for the case of the hole can be transferred to the case of the inclusion when the tips A and B are interchanged.

(ii) As when a hole is present, the crack extension angle for the case of an inclusion is largest when the crack is close to the inclusion. As the crack recedes from the inclusion this angle decreases and tends to zero as the crack moves to infinity.

(iii) When movement of the crack on horizontal lines close to the diameter of the inclusion occurs, the crack extends from tip B at a positive angle. However, as the distance c/a increases the crack extends from the tip A toward the inclusion for small values of b/a and away from the inclusion for greater values of b/a. As b/a further increases the crack extends from the tip B. Furthermore, when the crack moves in the direction of the applied stress and is completely overlapped by the inclusion, it always propagates away from the inclusion; while when the crack and the inclusion do not overlap the value of c/a determines whether the crack extends from tip A or B and toward or away from the inclusion.

(iv) The effect of the inclusion on the strength of the plate is opposite that of the hole. Thus, in all cases in which the presence of the hole decreases or increases the fracture strength of the plate the presence of the inclusion increases or decreases the fracture strength respectively. Therefore, all laws governing the dependence of the fracture stress on the geometrical configuration of the plate for the case of the hole can be transferred inversely to the case of the inclusion when the words 'increase' and 'decrease' are interchanged.

The results of this section were taken from reference [41].

References

[1] Williams, M.L., The stress around a fault or crack in dissimilar media, *Bulletin of the Seismological Society of America*, 49, pp. 199–204 (1959).

[2] Rice, J.R. and Sih, G.C., Plane problems of cracks in dissimilar media, *Journal of Applied Mechanics*, 32, pp. 418–423 (1965).

[3] Erdogan, F., Stress distribution in bonded dissimilar materials with cracks, *Journal of Applied Mechanics*, 32, pp. 403–410 (1965).

[4] England, A.H., A crack between dissimilar media, *Journal of Applied Mechanics*, 32, pp. 400–402 (1965).

[5] Malysev, B.M. and Salganik, R.L., The strength of adhesive joints using the theory of cracks, *International Journal of Fracture Mechanics*, 1, pp. 114–128 (1965).

[6] Sih, G.C., Hilton, P.D. and Wei, R.P., Exploratory development of fracture mechanics of composite systems, *Lehigh University Technical Report, AFML-TR-70-112* (1970).

[7] Sih, G.C., Hilton, P.D., Badaliance, R., Shenberger, P.S. and Villarreal, G., Fracture mechanics studies of composite systems, *Lehigh University Technical Report*, *AFML-TR-70-112*, Part II, June (1971).

[8] Sih, G.C., Hilton, P.D., Badaliance, R. and Villarreal, G., Exploratory development of fracture mechanics of composite systems, *Lehigh University Technical Report,* IFSM-73-28 (1973).

[9] Hilton, P.D. and Sih, G.C., A sandwiched layer of dissimilar materials weakened by crack-like imperfections, *Proceedings of the Fifth Southeastern Conference on Theoretical and Applied Mechanics*, edited by G.L. Rogers, S.C. Krane and E.G. Henneke, 5, pp. 123–149 (1970).

[10] Hilton, P.D. and Sih, G.C., A laminate composite with a crack normal to the interface, *International Journal of Solids and Structures*, 7, pp. 913–930 (1971).

[11] Hilton, P.D., Sih, G.C. and Badaliance, R., Analytic modeling of laminar composites with application to through crack problems, *Lehigh University Technical Report*, IFSM-73-53, (1973).

[12] Sih, G.C. and Chen, E.P., Fracture analysis of unidirectional composites, *Journal of Composite Materials*, 7, pp. 230–244 (1973).

[13] Chen, E.P. and Sih, G.C., Torsional and anti-plane strain delamination of an orthotropic layered composite, *Proceedings of the Thirteenth Mid-western Mechanics Conference*, 7, pp. 763–776 (1973).

[14] Chen, E.P. and Sih, G.C., Interfacial delamination of a layered composite under anti-plane strain, *Journal of Composite Materials*, 5, pp. 12–23 (1971).

[15] Wang, T.T., Kwei, T.K. and Zupko, H.M., Tensile strength of butt-joined epoxy-aluminum plates, *International Journal of Fracture Mechanics*, 6, pp. 127–137 (1970).

[16] Erdogan, F., Bonded dissimilar materials containing cracks parallel to the interface, *Engineering Fracture Mechanics*, 3, pp. 231–240 (1971).

[17] Erdogan, F. and Gupta, G.D., Layered composites with an interface flaw, *International Journal of Solids and Structures*, 7, pp. 1089–1107 (1971).

[18] Ashbaugh, N., Stress solution for a crack at an arbitrary angle to the interface, *International Journal of Fracture*, 11, pp. 205–219 (1975).

[19] Ratwani, M. and Gupta, G.D., Interaction between parallel cracks in layered composites, *Institut für Festkörpermechanik Report*, Freiburg, (1973).

[20] Ratwani, M. and Gupta, G.D., Interaction between parallel cracks in layered composites, *International Journal of Solids and Structures*, 10, pp. 701–708 (1974).

[21] Gupta, G.D., A layered composite with a broken laminate, *International Journal of Solids and Structures*, 9, pp. 1141–1154 (1973).

[22] Tamate, O. and Iwasaka, N., An arbitrarily oriented crack in a semi-infinite medium with a surface layer under tension, *International Journal of Solids and Structures*, 11, pp. 1257–1268 (1975).

[23] Cook, T.S. and Erdogan, F., Stresses in bonded materials with a crack perpendicular to the interface, *International Journal of Engineering Science*, 10, pp. 677–697 (1972).

[24] Leverenz, R.K., A finite element stress analysis of a crack in a bimaterial plate, *International Journal of Fracture Mechanics*, 8, pp. 317–324 (1972).

[25] Swenson, D.O. and Rau, C.A. Jr., The stress distribution around a crack perpendicular to an interface between materials, *International Journal of Fracture Mechanics*, 6, pp. 357–365 (1970).

[26] Lin, K.Y. and Mar, J.W., Finite element analysis of stress intensity factors for cracks at a bi-material interface, *International Journal of Fracture Mechanics*, 12, pp. 521-531 (1976).

[27] Fenner, D.N., Stress singularities in composite materials with an arbitrarily oriented crack meeting an interface, *International Journal of Fracture*, 12, pp. 705–721 (1976).

[28] Theocaris, P.S., Partly unbonded interfaces between dissimilar materials under normal and shear loading, *Acta Mechanica*, 24, pp. 99–115 (1976).

[29] Theocaris, P.S. and Gdoutos, E.E., Stress singularities in cracked composite full-planes, *International Journal of Fracture*, 11, pp. 205–219 (1975).

[30] Erdogan, F. and Biricikoglu, V., Two bonded half planes with a crack going through the interface, *Technical Report, NASA-TR-72-4* (1972).

[31] Erdogan, F. and Biricikoglu, V., Two bonded half planes with a crack going through the interface, NASA CR-2181 (1973).

[32] Erdogan, F. and Gupta, G.D., Inclusion problem with a crack crossing the boundary, *Lehigh University Technical Report*, IFSM-73-34 (1973).

[33] Erdogan, F., Gupta, G.D. and Ratwani, M., Interaction between a circular inclusion and an arbitrarily oriented crack, *Journal of Applied Mechanics*, 41, pp. 1007–1013 (1974).

[34] Perlman, A.B. and Sih, G.C., Elastostatic problems of curvilinear cracks in bonded dissimilar materials, *International Journal of Engineering Science*, 5, pp. 845–867 (1967).

[35] Sih, G.C., Fracture mechanics of adhesive joints, *Polymer Engineering and Science*, 20, pp. 977–981 (1980).

[36] Ioakimidis, N.I. and Theocaris, P.S., A system of curvilinear cracks in an isotropic elastic half-plane, *International Journal of Fracture*, 15, pp. 299–309 (1979).

[37] Gdoutos, E.E., Failure of a bimaterial plate with a crack at an arbitrary angle to the interface, *Fibre Science and Technology*, 15, pp. 27–40 (1981).

[38] Sih, G.C. and Macdonald, B., Fracture mechanics applied to engineering problems – strain energy density fracture criterion, *Engineering Fracture Mechanics*, 6, pp. 361–386 (1974).

[39] Gdoutos, E.E., Fracture of aluminum-epoxy layered composites containing cracks, *Journal of Strain Analysis*, 17, pp. 75–78 (1982).

[40] Sih, G.C., Prediction of crack growth under mixed mode conditions, *Proceedings of the US–Greece Seminar on Mixed Mode Crack Propagation*, Athens, Greece, August 18–22, 1980, Editors G.C. Sih and P.S. Theocaris, Sijthoff and Noordhoff, pp. 3–19 (1981).

[41] Gdoutos, E.E., Interaction effects between a crack and a circular inclusion, *Fibre Science and Technology*, 15, pp. 173–185 (1981).

8

Plates and shells

8.1. Introduction

While in the previous chapters two-dimensional crack problems under various geometrical configurations and loading conditions were studied, the present chapter extends the strain energy density criterion to analyze the brittle fracture behavior of cracked bent plates and shells. A plate is defined as a body whose dimension in the third direction is small compared to its other two dimensions and is usually loaded perpendicularly to its main plane. Therefore, bending stresses in a plate are predominantly developed. A shell also extends mainly along two dimensions but its middle surface is generally curved (cylindrical, spheroidal, etc.). The predominant deformation of a shell consists of bending stresses in combination with in-plane membrane stresses.

The theory of the strain energy density criterion developed in chapter one is applied to the problems of bent plates and shells. The basic problem of a plate with an inclined crack under the direction of the applied uniaxial moment is studied first. This problem corresponds to the case of a plate with an inclined crack loaded by a uniaxial tensile or compressive stress studied in Chapter 2 (page 13). The basic equations used to determine the crack extension angle from the most vulnerable layer of the plate as well as the critical value of the applied moment are established. Thus the dependence of these quantities on the crack inclination angle, the Poisson's ratio and the thickness of the plate is demonstrated. Second, the case of a pressurized vessel consisting of a long cylindrical shell with two hemispherical heads is investigated. The critical value of the applied pressure for extension of a crack in the cylindrical or semispherical area of the vessel is determined. In addition, the values of the angle formed by the fracture path with the initial crack direction are determined.

8.2. A cracked bent plate with an inclined crack

a. *The local stress field*: The cracked plate of thickness h shown in Figure 8.1 is referred to a system of Cartesian coordinates with the x, y-plane coinciding with the middle plane (the plane that bisects the plate thickness) of the plate. Conventional plate theories assume the stress variation through the thickness of the plate is known as *a priori*. This hypothesis has been disproven in the work by Hartranft and Sih [1] who presented a more rigorous plate theory. The stress component σ_z of a thin plate can be assumed to be zero throughout the plate. The in-plane stresses σ_x, σ_y and τ_{xy} on the element located at (x, y, z) vary linearly with z (assuming straight lines normal to the mid-plane of the plate remain straight after deformation and normal to the deformed middle plane, analogous to Bernulli's hypothesis for long beams), i.e.,

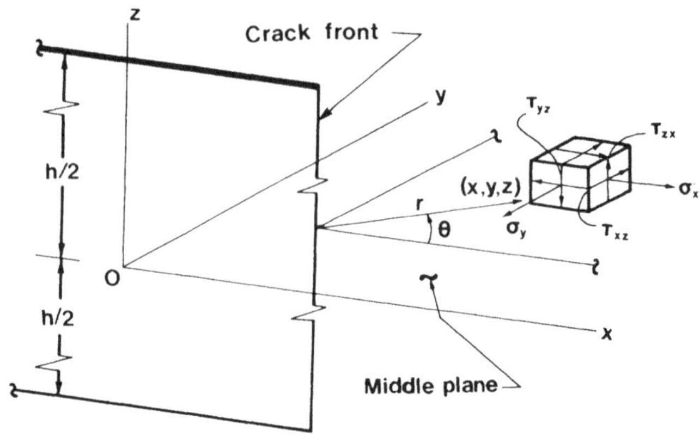

Figure 8.1. Stresses on an element ahead of crack in a thin plate.

$$\sigma_x = \frac{12z}{h^3} M_x \qquad \sigma_y = \frac{12z}{h^3} M_y \qquad \tau_{xy} = \frac{12z}{h^3} M_{xy} \tag{8.1}$$

where M_x and M_y are the bending moments and M_{xy} the twisting moment per unit length acting on any section of the plate parallel to the xz and yz planes. These values can be obtained by integration:

$$M_x(x, y) = \int_{-h/2}^{h/2} \sigma_x(x, y, z) z \, dz \tag{8.2a}$$

$$M_y(x, y) = \int_{-h/2}^{h/2} \sigma_y(x, y, z) z \, dz \tag{8.2b}$$

$$M_{xy}(x, y) = \int_{-h/2}^{h/2} \tau_{xy}(x, y, z) z \, dz. \tag{8.2c}$$

The surface conditions at $z = \pm h/2$ are satisfied by requiring that:

$$\tau_{xz}(x, y, \pm h/2) = \tau_{yz}(x, y, \pm h/2) = 0 \tag{8.3}$$

where the normal stress σ_z is assumed to be zero throughout the plate, as noted above.

From equilibrium equations it is deduced that shearing stresses τ_{xz} and τ_{yz} vary parabolically with z as:

$$\tau_{xz} = \frac{3}{2h}\left[1 - \left(\frac{2z}{h}\right)^2\right] V_x \tag{8.4a}$$

$$\tau_{yz} = \frac{3}{2h}\left[1 - \left(\frac{2z}{h}\right)^2\right] V_y \tag{8.4b}$$

where V_x and V_y are the shearing forces per unit length on sections parallel to the y and x axes, given by

$$V_x(x, y) = \int_{-h/2}^{h/2} \tau_{xz}(x, y, z) \, dz \tag{8.5a}$$

$$V_y(x, y) = \int_{-h/2}^{h/2} \tau_{yz}(x, y, z)\, dz. \tag{8.5b}$$

Relations (8.1) and (8.4) establish the z-dependence of the stress components of the plate. Thus, the plate problem has been essentially reduced to two dimensions and the boundary conditions will be specified in terms of the moments and shear forces.

Earlier investigations [2–4] used the classical Poisson–Kirchhoff theory to determine the bending stresses near the crack ends. This theory neglects the effect of transverse shear deformation, reducing the number of boundary conditions satisfied along the crack face below that which is found in reality. This reduction in boundary conditions leads to an erroneous stress distribution in the neighborhood of the crack tip. It is thus found that the angular variation of the stresses is a function of the Poisson's ratio of the plate, and the stresses τ_{xz} and τ_{yz} have a singularity of the order $r^{-3/2}$. These results contrast with the exact three-dimensional solution of the problem [5–6]. Thus, the approximate Kirchhoff conditions are inadequate for determination of the singular stresses in the neighborhood of the crack.

The inaccuracies caused by the Kirchhoff conditions were resolved by use of the Reissner's [7, 8] sixth order bending theory proposed by Knowles and Wang [9] and Hartranft and Sih [5]. The Reissner theory admits the following three boundary conditions for a crack placed on the x-axis:

$$M_y(x, 0) = M_{xy}(x, 0) = V_y(x, 0) = 0 \tag{8.6}$$

satisfied on the segment of the x-axis where the crack is located. Thus, the following expressions for the moments and shear forces at the crack tip are obtained:

$$\begin{aligned} M_x = {} & \frac{K_1}{(2r)^{1/2}} \cos\frac{\theta}{2}\left(1 - \sin\frac{\theta}{2}\sin\frac{3\theta}{2}\right) \\ & - \frac{K_2}{(2r)^{1/2}} \sin\frac{\theta}{2}\left(2 + \cos\frac{\theta}{2}\cos\frac{3\theta}{2}\right) + O(1) \end{aligned} \tag{8.7a}$$

$$\begin{aligned} M_y = {} & \frac{K_1}{(2r)^{1/2}} \cos\frac{\theta}{2}\left(1 + \sin\frac{\theta}{2}\sin\frac{3\theta}{2}\right) \\ & + \frac{K_2}{(2r)^{1/2}} \sin\frac{\theta}{2}\cos\frac{\theta}{2}\cos\frac{3\theta}{2} + O(1) \end{aligned} \tag{8.7b}$$

$$\begin{aligned} M_{xy} = {} & \frac{K_1}{(2r)^{1/2}} \sin\frac{\theta}{2}\cos\frac{\theta}{2}\cos\frac{3\theta}{2} \\ & + \frac{K_2}{(2r)^{1/2}} \cos\frac{\theta}{2}\left(1 - \sin\frac{\theta}{2}\sin\frac{3\theta}{2}\right) + O(1) \end{aligned} \tag{8.7c}$$

$$V_x = -\frac{K_3}{(2r)^{1/2}} \sin\frac{\theta}{2} + O(1) \tag{8.7d}$$

$$V_y = \frac{K_3}{(2r)^{1/2}} \cos\frac{\theta}{2} + O(1) \tag{8.7e}$$

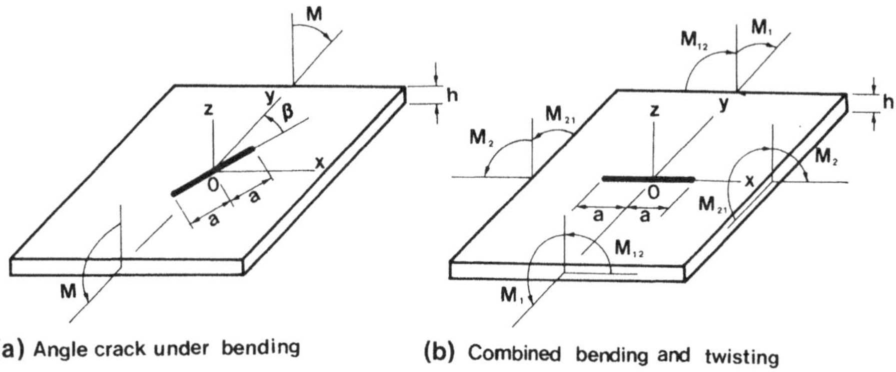

Figure 8.2. Loading configuration for bending and twisting of a cracked plate.

where K_1 and K_2 are the moment intensity factors and K_3 the shear force intensity factor. Note that the $r^{-1/2}$ singularity and the θ-dependence in equations (8.7) are the same as for a crack subjected to in-plane loads under mixed-mode loading conditions (Chapter 2, equations (2.1)). This is consistent with the view that each layer of the plate experiences extension or compression, with the aggregate of layers producing bending.

Consider now the case of a through crack in a bent plate subjected to a bending moment of magnitude M whose plane makes an angle β with the crack plane (Figure 8.2a). This loading is equivalent to the application of bending moments M_1, M_2 and twisting moment M_{12} (Figure 8.2b) given according to the transformation properties of stresses by:

$$M_1 = M \sin^2\beta \tag{8.8a}$$

$$M_2 = M \cos^2\beta \tag{8.8b}$$

$$M_{12} = M \sin\beta \cos\beta. \tag{8.8c}$$

For this problem, Hartranft and Sih [10] and Wang [11] gave the following expressions for the moment K_1, K_2 and shear force K_3 intensity factors:

$$K_1 = \Phi(1) M_1 \sqrt{a} \tag{8.9a}$$

$$K_2 = \Psi(1) M_{12} \sqrt{a} \tag{8.9b}$$

$$K_3 = -\frac{\sqrt{10}}{(1+\nu)h} \Omega(1) M_{12} \sqrt{a}. \tag{8.9c}$$

The separate effects of M_1, M_2 and M_{12} are illustrated in Figures 8.3(a), (b) and (c), where normal bending produces K_1, parallel bending produces no stresses and pure twisting leads to both K_2 and K_3. The functions $\Phi(1)$, $\Psi(1)$ and $\Omega(1)$ are computed numerically from integral equations [10, 11]. Their values as a function of $h/a\sqrt{10}$ for various Poisson's ratios are found in Figures 8.4 to 8.6.

Substituting relations (8.7) into (8.1) and (8.4), and taking into account relations (8.8) and (8.9), we obtain the following expressions for the stress components σ_x, σ_y, τ_{xy}, τ_{xz} and τ_{yz}:

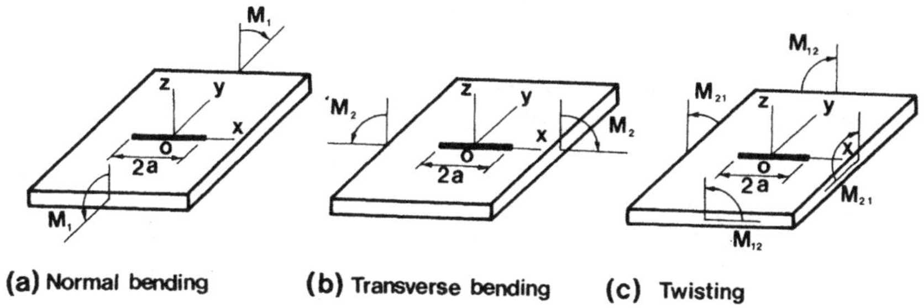

Figure 8.3. Three separate problems of plate bending.

$$\sigma_x = \frac{k_1}{(2r)^{1/2}}\cos\frac{\theta}{2}\left(1-\sin\frac{\theta}{2}\sin\frac{3\theta}{2}\right)-\frac{k_2}{(2r)^{1/2}}\sin\frac{\theta}{2}\left(2+\cos\frac{\theta}{2}\cos\frac{3\theta}{2}\right)+O(1) \tag{8.10a}$$

$$\sigma_y = \frac{k_1}{(2r)^{1/2}}\cos\frac{\theta}{2}\left(1+\sin\frac{\theta}{2}\sin\frac{3\theta}{2}\right)+\frac{k_2}{(2r)^{1/2}}\sin\frac{\theta}{2}\cos\frac{\theta}{2}\cos\frac{3\theta}{2}+O(1) \tag{8.10b}$$

$$\tau_{xy} = \frac{k_1}{(2r)^{1/2}}\sin\frac{\theta}{2}\cos\frac{\theta}{2}\cos\frac{3\theta}{2}+\frac{k_2}{(2r)^{1/2}}\cos\frac{\theta}{2}\left(1-\sin\frac{\theta}{2}\sin\frac{3\theta}{2}\right)+O(1) \tag{8.10c}$$

$$\tau_{xz} = -\frac{k_3}{(2r)^{1/2}}\sin\frac{\theta}{2}+O(1) \tag{8.10d}$$

$$\tau_{yz} = \frac{k_3}{(2r)^{1/2}}\cos\frac{\theta}{2}+O(1) \tag{8.10e}$$

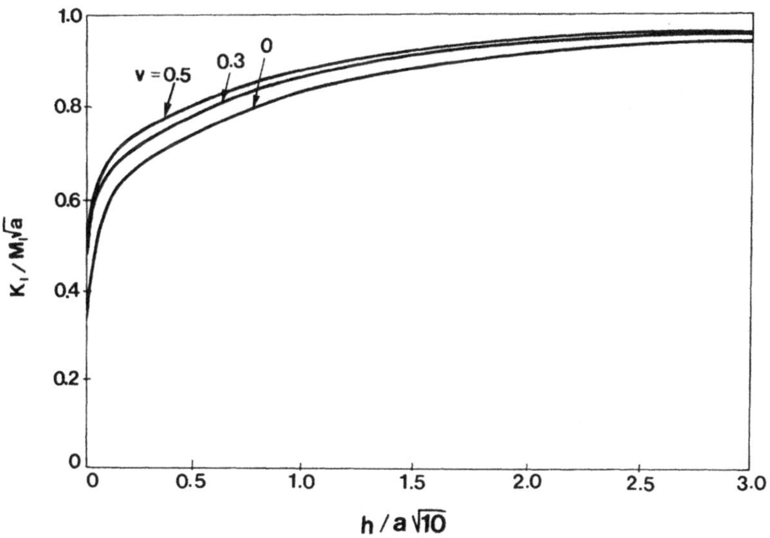

Figure 8.4. Bending-moment-intensity factor versus ratio of thickness to crack length.

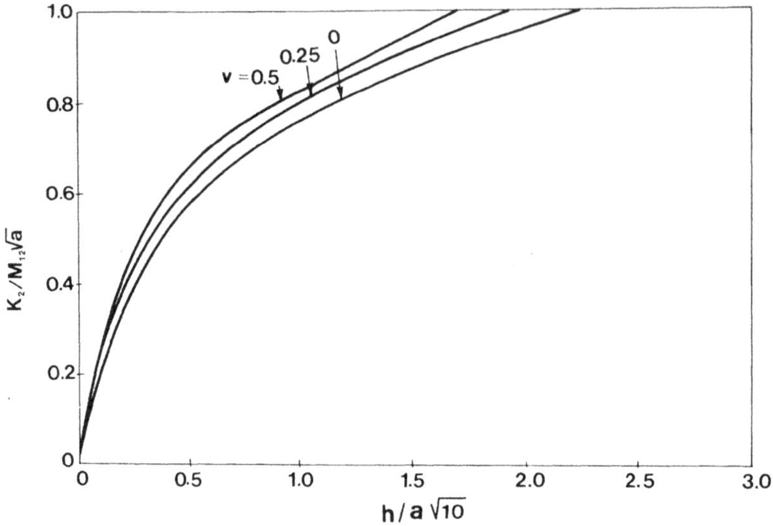

Figure 8.5. Twisting moment-intensity factor versus ratio of thickness to crack length.

where the coefficients k_j ($j = 1, 2, 3$) (called stress intensity factors) are related to K_j ($j = 1, 2, 3$) by

$$k_1 = \frac{12z}{h^3} K_1 \tag{8.11a}$$

$$k_2 = \frac{12z}{h^3} K_2 \tag{8.11b}$$

$$k_3 = \frac{3}{2h}\left[1 - \left(\frac{2z}{h}\right)^2\right] K_3 \tag{8.11c}$$

and are given by:

$$k_1 = \frac{12z}{h^3} \Phi(1) M\sqrt{a} \sin^2\beta \tag{8.12a}$$

$$k_2 = \frac{12z}{h^3} \Psi(1) M\sqrt{a} \sin\beta\cos\beta \tag{8.12b}$$

$$k_3 = -\frac{3\sqrt{10}}{2(1+\nu)h^2}\left[1 - \left(\frac{2z}{h}\right)^2\right] \Omega(1) M\sqrt{a} \sin\beta\cos\beta. \tag{8.12c}$$

b. *Equations of the strain energy density criterion*: Introducing the values of stress components σ_x, σ_y, τ_{xy}, τ_{xz} and τ_{yz} from relations (8.10) into (1.2) and (1.3) we obtain for the strain energy density factor S the form:

$$S = A_{11}k_1^2 + 2A_{12}k_1k_2 + A_{22}k_2^2 + A_{33}k_3^2 \tag{8.13}$$

where the coefficients A_{ij} ($i, j = 1, 2, 3$) are given by:

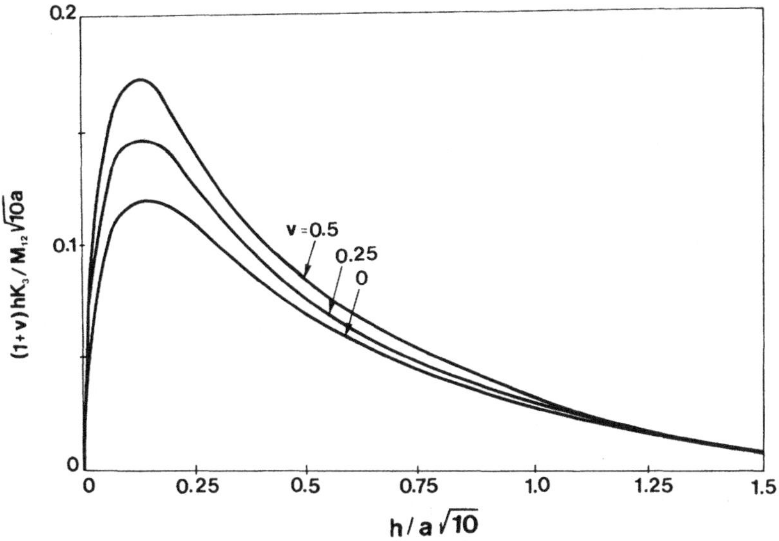

Figure 8.6. Shear intensity factor versus ratio of thickness to crack length.

$$A_{11} = \frac{1}{8E}(1 + \cos\theta)[3 - \nu - (1 + \nu)\cos\theta] \tag{8.14a}$$

$$A_{12} = \frac{1}{4E}\sin\theta[(1 + \nu)\cos\theta - (1 - \nu)] \tag{8.14b}$$

$$A_{22} = \frac{1}{8E}[4(1 - \cos\theta) + (1 + \nu)(3\cos\theta - 1)(1 + \cos\theta)] \tag{8.14c}$$

$$A_{33} = \frac{1 + \nu}{2E}. \tag{8.14d}$$

Note that formula (8.10) describing the stress components for the bent plate problem is identical to formula (2.1), which corresponds to a description of stress components for the general two-dimensional problem under generalized plane stress conditions ($\kappa = (3 - \nu)/(1 + \nu)$). As a result, the value of S given by equations (8.13) and (8.14) with $k_3 = 0$ coincides with its corresponding value for the extensional problem (equations (2.3) and (2.4)). The values, however, of the stress intensity factors for the two cases differ in that the stress intensity factors for the plate problem vary along the crack front, while remaining constant for the extensional problem.

Introducing the values of the stress intensity factors k_1, k_2 and k_3 given by relations (8.12) into equation (8.13) we obtain the following expression for the strain energy density factor S:

$$S = \frac{2a}{E}\left[\frac{3\Phi(1)M\sin\beta}{h^2}\right]^2\left\{\left(\frac{z}{h}\right)^2 F(\theta, \beta) + \left[1 - \left(\frac{2z}{h}\right)^2\right]^2 G(\beta)\right\} \tag{8.15}$$

with the functions $F(\theta, \beta)$ and $G(\beta)$ defined as

$$F(\theta,\beta) = B_{11}\sin^2\beta + 2\lambda B_{12}\sin\beta\cos\beta + \lambda^2 B_{22}\cos^2\beta \tag{8.16a}$$

$$G(\beta) = \frac{5}{8(1+\nu)}\gamma^2\cos^2\beta \tag{8.16b}$$

and the coefficients B_{ij} and λ and γ given by:

$$B_{ij} = 8EA_{ij} \quad (i,j = 1,2,3) \tag{8.17}$$

$$\lambda = \frac{\Psi(1)}{\Phi(1)} \qquad \gamma = \frac{\Omega(1)}{\Phi(1)}. \tag{8.18}$$

The numerical values of λ and γ can be obtained from the graphs of Figures 8.4 to 8.6.

One of the basic assumptions of the strain energy density criterion is the requirement of the stationary value of the strain energy density factor S (equation (1.10)). This is expressed by the following equation:

$$\begin{aligned}&(1+\nu)\{2(1-3\lambda^2)\sin 2\theta_0 - (1-\lambda)(1-3\lambda)\sin[2(\theta_0+\beta)]\\&\quad -(1+\lambda)(1+3\lambda)\sin[2(\theta_0-\beta)]\} - 2(1-\nu)[2(1-\lambda^2)\sin\theta_0\\&\quad -(1-\lambda)^2\sin(\theta_0+2\beta) - (1+\lambda)^2\sin(\theta_0-2\beta)] = 0.\end{aligned} \tag{8.19}$$

c. *Direction of crack extension*: Assuming a cracked plate subjected to a uniformly distributed moment whose plane makes an angle of inclination β with respect to the crack direction, solution of equation (8.19) under the restrictions imposed by inequality (1.11) gives the values of the crack extension angle θ_0. Of the two values of θ_0 which meet these requirements, one is negative, corresponding to the tension side of the plate, and the other is positive, corresponding to the compression side. In most materials, tension strength is smaller than compression strength. Thus the crack initiates from the tension side and therefore only negative fracture angles will be considered.

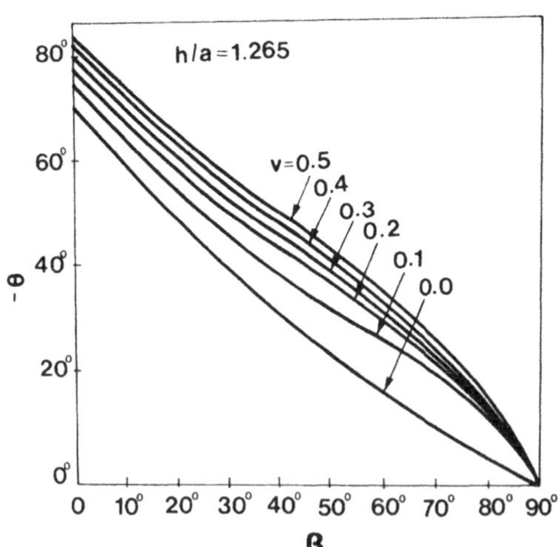

Figure 8.7. Variations of fracture angle with crack angle for constant h/a.

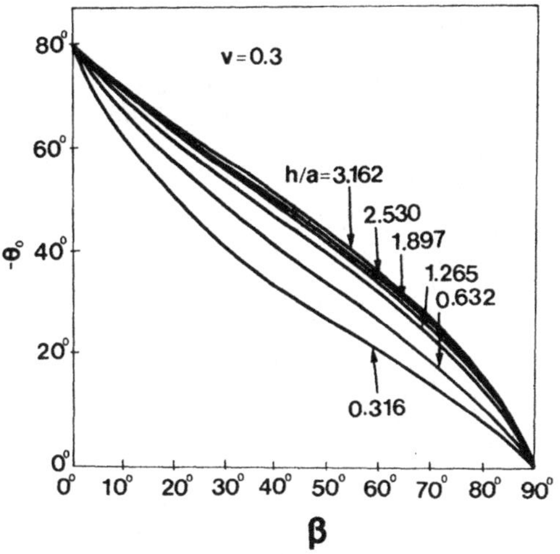

Figure 8.8. Variations of fracture angle with crack angle for constant ν.

Figure 8.7 presents the variation of the angle $-\theta_0$ versus the crack inclination angle β for $h/a = 1.265$. The various curves of the figure correspond to the values of the Poisson's ratio ν of the plate material equal to 0, 0.1, 0.2, 0.3, 0.4 and 0.5. This figure shows that for $\beta = 90°$ the crack extension angle θ_0 is equal to zero. In other words, when the direction of the plane of the applied moment is perpendicular to the crack, the crack propagates in its own plane. As the crack inclination angle β increases the fracture angle $-\theta_0$ also increases, so that the direction of crack initiation deviates more and more from its initial plane. We also see from Figure 8.7 that, the angle $-\theta_0$ is substantially influenced by the Poisson's ratio of the material, especially, when the angle β is in the middle range.

Unlike the extensional case, the ratio h/a of plate thickness to half-crack length influences the direction of crack growth. The variation of the angle $-\theta_0$ versus β for $\nu = 0.3$ and various values of the ratio h/a is displayed in Figure 8.8, where it is seen that $-\theta_0$ decreases as the plate thickness increases.

d. *Critical moments*: Introducing the previously obtained values of the crack extension angle θ_0 into equation (8.15), the minimum values $S_{\min}$ of the strain energy density factor are determined. When $S_{\min}$ is then equated to the critical value S_{cr} of S, which was proved to be a material constant, the critical values M_{cr} of the applied moment M for crack extension are obtained. Thus, the following relation is derived from relation (8.15)

$$\frac{M_{cr}}{(Eh^3S_{cr})^{1/2}} = \frac{\sqrt{h/a}}{3\sqrt{2}\phi(1)\sin\beta\{(z/h)^2F_{\min}(\theta_0,\beta)+[1-(2z/h)^2]G(\beta)\}^{1/2}} \quad (8.20)$$

where $F_{\min}(\theta_0, \beta)$ is the minimum value of the function $F(\theta, \beta)$ which is the only quantity in relation (8.15) dependent on θ. Relation (8.20) gives the normalized critical moment $M_{cr}/(Eh^3S_{cr})^{1/2}$ as a function of the crack inclination angle β, the ratio of the plate thickness to the half-crack length h/a, the Poisson's ratio ν and the ratio z/h of the distance z of the particular layer considered to the plate thickness.

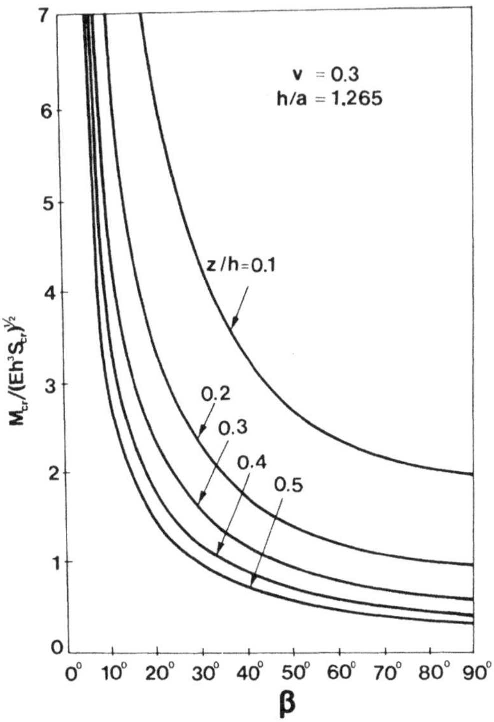

Figure 8.9. Normalized moment versus crack angle for constant ν and h/a.

In order to find the particular layer of the plate from which crack extension takes place let us study the dependence of M on z for fixed values of ν and h/a. Figure 8.9 presents the variation of $M_{cr}/(Eh^3S_{cr})^{1/2}$ versus β for $\nu = 0.3$ and $h/a = 1.265$ for different values of z/h from 0.1 to 0.5. The curve for $z/h = 0$ lies beyond the scale range of the figure. Each value of z/h corresponds to a material layer of the plate with $z = 0$ being the middle layer and $z/h = 0.5$ the surface layer. It is observed that the critical moment M_{cr} for crack extension is always smaller at the surface layer and therefore fracture of the plate always takes place from the surface layer. Thus for $z/h = 0.5$ Figure 8.10 presents the variation of $M_{cr}/(Eh^3S_{cr})^{1/2}$ versus the crack inclination angle β for $h/a = 1.265$. The various curves of the figure correspond to the values of the Poisson's ratio ν equal to 0, 0.1, 0.2, 0.3, 0.4 and 0.5. Figure 8.11 establishes the dependence of the critical moment on the ratio h/a of the plate thickness to the half-crack length for a fixed value of the Poisson's ratio ν equal to 0.3. The curves correspond to the values of h/a equal to 0.316, 0.632, 1.265, 1.897, 2.530 and 3.162. Figures 8.10 and 8.11 indicate that M_{cr} decreases with the crack inclination angle β (similar to results obtained for the inclined crack under extensional loading, Section 2.3, page 13), while it increases with ν and h/a. Thus, when the thickness of the plate increases the value of the critical moment causing fracture also increases.

The results of the present section were obtained from reference [12].

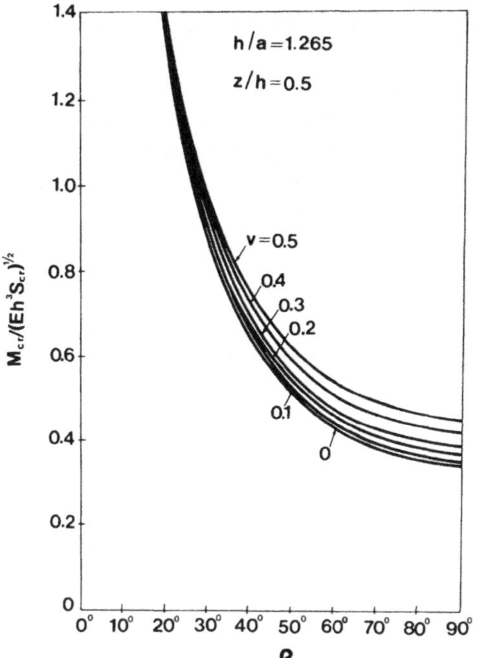

Figure 8.10. Normalized moment versus crack angle for constant h/a and z/h.

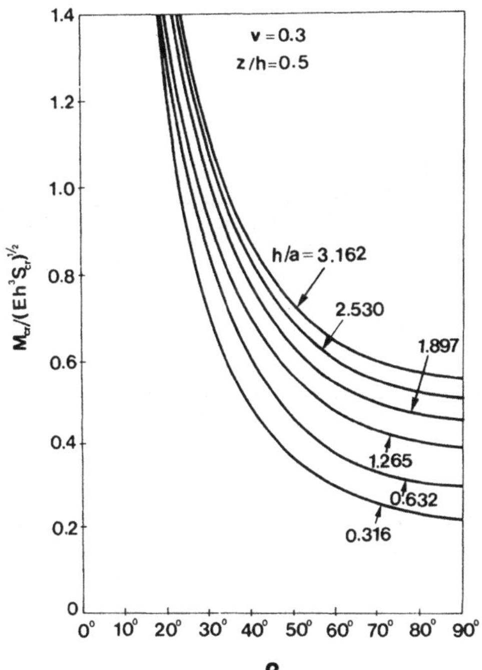

Figure 8.11. Normalized moment versus crack angle for constant ν and z/h.

8.3. A cracked cylindrical shell with semispherical heads

In the present section the strain energy density criterion will be used to determine the critical loads of a cracked cylindrical shell with semispherical heads subjected to internal pressure. Two cases will be studied below in detail; (a) A crack of any orientation in the cylindrical shell and (b) a crack of any orientation in the hemispherical heads. In each case the crack will be considered to be located in the membrane and also in the non-membrane area.

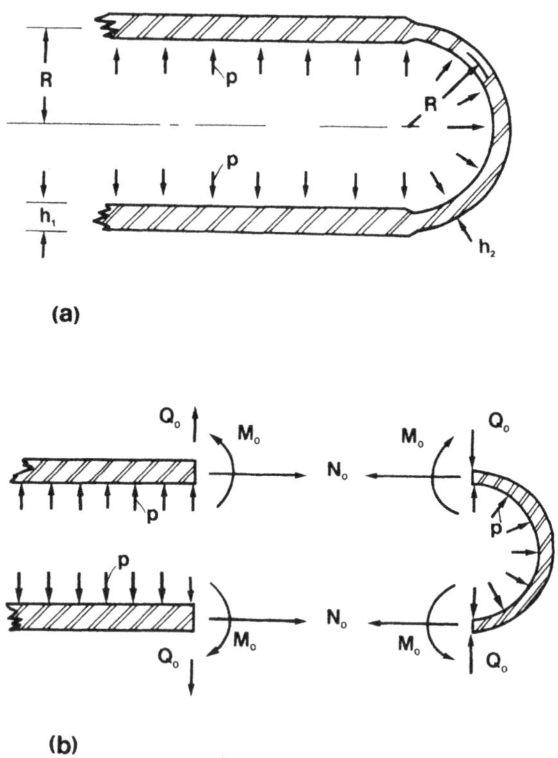

Figure 8.12. Geometry (a) and free-body diagram (b) of a half of a cylindrical shell with semispherical heads.

a. *The stress distribution*: Consider the shell of Figure 8.12a consisting of a long cylindrical part with a semispherical head both of radius R. Let h_1 and h_2 be the thicknesses of the cylindrical and semispherical parts respectively. The free-body diagram showing the forces and moments at the shell juncture is shown in Figure 8.12b. p is the value of the applied internal stress and Q_0 and M_0 the shear force and bending moment at the shell juncture. The meridional stress σ_ξ and the hoop stress σ_θ at a distance ξ from the juncture will be given by the following relations [13]

$$\sigma_\xi = \frac{pR}{2h_i} \pm \frac{6M_\xi}{h_i^2} \tag{8.21a}$$

$$\sigma_\theta = \frac{pR}{k_i h_i} + \frac{N_\theta}{h_i} + \frac{6M_\theta}{h_i^2} \tag{8.21b}$$

where

$$M_\xi = \frac{Q_0}{B_i} e^{-B_i\xi} \sin B_i\xi + M_0\, e^{-B_i\xi}(\sin B_i\xi + \cos B_i\xi) \tag{8.22a}$$

$$M_\theta = \nu M_\xi \tag{8.22b}$$

$$N_\theta = 2Q_0 B_i R\, e^{-B_i\xi} \cos B_i\xi + 2M_0 B_i^2 R\, e^{-B_i\xi}(\cos B_i\xi - \sin B_i\xi) \tag{8.22c}$$

$$B_i^4 = \frac{3(1-\nu^2)}{R^2 h_i^2} \tag{8.22d}$$

$$D_i = \frac{Eh_i^3}{12(1-\nu^2)} \tag{8.22e}$$

with the quantities Q_0 and M_0 given by [14]:

$$\frac{Q_0}{\dfrac{pR^2B_1^3D_1}{Eh_2}\left[(2-\nu)\left(\dfrac{h_2}{h_1}-1\right)+1\right]} = \frac{1+\left(\dfrac{h_1}{h_2}\right)^{5/2}}{\left[1+\left(\dfrac{h_1}{h_2}\right)^{5/2}\right]\left[1+\left(\dfrac{h_1}{h_2}\right)^{3/2}\right]-\dfrac{1}{2}\left[1-\left(\dfrac{h_1}{h_2}\right)^2\right]^2} \tag{8.23a}$$

$$\frac{M_0}{\dfrac{pR^2B_1^2D_1}{2Eh_2}\left[(2-\nu)\left(\dfrac{h_2}{h_1}-1\right)+1\right]} = \frac{1-\left(\dfrac{h_1}{h_2}\right)^2}{\left[1+\left(\dfrac{h_1}{h_2}\right)^{5/2}\right]\left[1+\left(\dfrac{h_1}{h_2}\right)^{3/2}\right]-\dfrac{1}{2}\left[1-\left(\dfrac{h_1}{h_2}\right)^2\right]^2}\,. \tag{8.23b}$$

In the above relations, k_i is equal to 1 for the cylindrical shell, and 2 for the hemispherical shell. E is the modulus of elasticity, and ν the Poisson's ratio. The upper signs in equations (8.21) refer to the inner surfaces of the shells, and the lower signs to the outer surfaces.

The previous developments show that an element of the shell in either the cylindrical or the hemispherical regions is generally subjected to membrane and bending stresses. Figure 8.13 illustrates a crack of length $2a$ subjected to the normal extensional

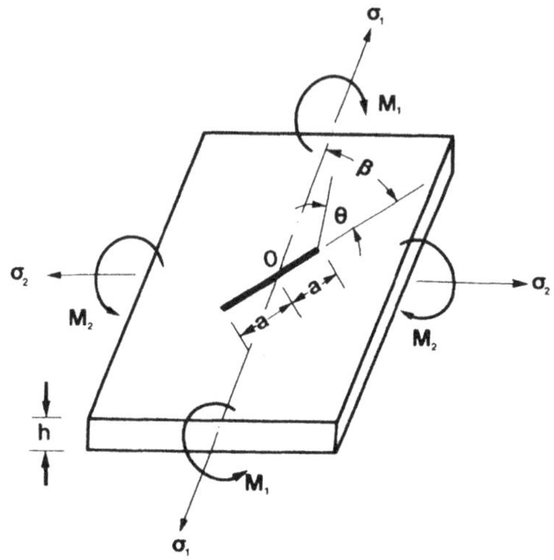

Figure 8.13. An inclined crack subjected to combined biaxial in-plane stresses and bending moments.

stresses σ_1, σ_2 and the bending moments M_1, M_2. The crack is inclined at an angle β to the direction of the stress σ_1 (or the plane of the moment M_1).

Considering the biaxial extensional stresses and bending moments we obtain the following relations for k_1 and k_2 by superimposing relations (2.7) and (8.12):

$$k_1 = \left[\frac{12z}{h^3}\Phi(1)(M_1\sin^2\beta + M_2\cos^2\beta) + \sigma_1\cos^2\beta + \sigma_2\sin^2\beta\right]\sqrt{a} \tag{8.24a}$$

$$k_2 = \left[\frac{12z}{h^3}\Psi(1)(M_1 - M_2)\sin\beta\cos\beta + (\sigma_1 - \sigma_2)\sin\beta\cos\beta\right]\sqrt{a} \tag{8.24b}$$

$$k_3 = -\frac{3\sqrt{10}}{2(1+\nu)h^2}\left[1 - \left(\frac{2z}{h}\right)^2\right]\Omega(1)(M_1 - M_2)\sin\beta\cos\beta\sqrt{a}. \tag{8.24c}$$

b. *The cylindrical shell*: A crack of length $2a$ and of inclination angle β with respect to the meridional direction is embedded in the cylindrical shell. When the crack is far from the shell juncture, membrane stresses predominate in the shell. $M_1 = M_2 = 0$ and the membrane stresses σ_1 and σ_2 are given from (8.21) by:

$$\sigma_1 = \sigma_\theta = \frac{pR}{h} \tag{8.25a}$$

$$\sigma_2 = \sigma_\xi = \frac{pR}{2h}. \tag{8.25b}$$

On the other hand when the crack is in the non-membrane region (close to the shell juncture) the effect of bending stresses must also be taken into account. Numerical results are obtained by considering an element at a distance $\xi = 4$ in from the shell juncture, when $h_1 = h_2 = 3/8$ in and $R = 60$ in. These values allow

Table 8.1. Fracture angle versus crack angle and crack length for cylindrical shell ($z/h = 0.5$; $\nu = 0.3$).

$2a/h$ \ β	0°	10°	20°	30°	40°	50°	60°	70°	80°	90°
	Fracture angle, $-\theta_0$									
1	0	14.03	22.59	26.35	26.91	25.14	21.44	15.90	8.57	0
2	0	14.67	23.44	27.22	27.74	25.92	22.14	16.45	8.89	0
3	0	15.02	23.89	27.68	28.18	26.33	22.51	16.76	9.07	0
4	0	15.25	24.18	27.98	28.47	26.60	22.75	16.95	9.19	0
5	0	15.40	24.38	28.18	28.66	26.78	22.91	17.08	9.27	0
6	0	15.51	24.52	28.32	28.80	26.91	23.03	17.18	9.33	0
7	0	15.59	24.63	28.44	28.91	27.01	23.12	17.26	9.37	0
8	0	15.68	24.74	28.54	29.01	27.10	23.20	17.32	9.41	0
9	0	15.74	24.82	28.62	29.08	27.17	23.27	17.37	9.44	0
10	0	15.79	24.88	28.68	29.14	27.23	23.32	17.41	9.47	0
–	0	17.44	26.74	30.28	30.40	28.16	23.98	17.85	9.69	0

computation of the extensional stresses $\sigma_{1,2}$, the bending moments $M_{1,2}$ and consequently the stress intensity factors k_1, k_2 and k_3 (relations (8.24)). At these values of k_1, k_2 and k_3 and assuming plane strain conditions predominant in the shell, the strain energy density factor S is calculated from relation (8.13). The values of A_{11}, A_{12}, A_{22} and A_{33} are given by (putting $\kappa = 3 - 4\nu$ into relations (2.3)):

$$16\mu A_{11} = (1 + \cos\theta)(3 - 4\nu - \cos\theta) \tag{8.26a}$$

$$16\mu A_{12} = 2\sin\theta[\cos\theta - (1 - 2\nu)] \tag{8.26b}$$

$$16\mu A_{22} = 4(1 - \nu)(1 - \cos\theta) + (1 + \cos\theta)(3\cos\theta - 1) \tag{8.26c}$$

$$4\mu A_{33} = 1. \tag{8.26d}$$

The values of the crack extension angle $-\theta_0$ are obtained from the stationary (minimum) values of S, while restricting attention to the more vulnerable surface layer of the shell ($z/h = 0.5$).

These values are presented in Table 8.1 as a function of the crack angle β and the crack length to the plate thickness ratio $2a/h$. The first ten values of $-\theta_0$ corresponding to each particular value of $2a/h$ are referred to the non-membrane region, while the last row is independent of the crack length, and corresponds to the membrane region of the cylindrical shell. Using these values of $-\theta_0$ the corresponding values of the non-dimensional quantity $\bar{S}_{cr}$ defined by:

$$\bar{S}_{cr} = \frac{64\mu S_{cr}}{a\sigma_1^2} \tag{8.27}$$

are obtained. These values are displayed in Table 8.2.

Substituting the value of S_{cr} in terms of the critical value of the stress intensity factor k_1 given by relation (2.16) into relation (8.27) we obtain:

$$\bar{S}_{cr} = \frac{16(1 - 2\nu)}{a}\left(\frac{k_{1c}}{\sigma_1}\right)^2. \tag{8.28}$$

Table 8.2. Values of $\bar{S}_{cr}$ for cylindrical shell ($z/h = 0.5$; $\nu = 0.3$)

$2a/h$ \ β	0°	10°	20°	30°	40°	50°	60°	70°	80°	90°
1	2.20	2.35	2.76	3.40	4.16	4.96	5.69	6.29	6.68	6.81
2	2.14	2.29	2.71	3.36	4.13	4.93	5.67	6.26	6.64	6.78
3	2.11	2.26	2.69	3.34	4.12	4.93	5.66	6.24	6.63	6.76
4	2.09	2.24	2.67	3.33	4.12	4.92	5.65	6.24	6.62	6.75
5	2.08	2.23	2.66	3.33	4.11	4.92	5.65	6.23	6.61	6.75
6	2.07	2.22	2.66	3.32	4.11	4.92	5.65	6.23	6.61	6.74
7	2.07	2.21	2.65	3.32	4.11	4.91	5.64	6.23	6.61	6.74
8	2.06	2.21	2.65	3.31	4.10	4.91	5.64	6.22	6.60	6.73
9	2.05	2.20	2.64	3.31	4.10	4.91	5.64	6.22	6.60	6.73
10	2.05	2.20	2.64	3.31	4.10	4.91	5.63	6.22	6.59	6.73
–	1.60	1.75	2.18	2.85	3.65	4.47	5.23	5.85	6.26	6.40

For a shell made of D6AC steel with $E = 30 \times 10^6$ psi, $\nu = 0.3$ and $K_{1c} = \sqrt{\pi}k_1 = 80\,\mathrm{ksi}\sqrt{\mathrm{in}}$, and with values of the geometrical dimensions of the shell and an element at a distance ξ from the shell juncture given above, we obtain for the critical pressure p_{cr} the following values:

$$p_{cr} = \frac{0.7136}{\sqrt{a\bar{S}_{cr}}}\,\mathrm{ksi} \tag{8.29}$$

for the membrane region, and

$$p_{cr} = \frac{0.6865}{\sqrt{a\bar{S}_{cr}}}\,\mathrm{ksi} \tag{8.30}$$

for the non-membrane region. The variation of p_{cr} versus the crack length $2a$ for various values of the crack inclination angle β is given by Figure 8.14 for the membrane and by Figure 8.15 for the non-membrane region. These figures illustrate the trend of decreasing critical pressure with increasing crack length and decreasing crack inclination angle, which was previously established for the extensional and bending problems. Furthermore, we see that for the same crack length and crack angle the critical pressure is higher for the membrane region. It is also observed that the critical pressure is more sensitive to the crack-angle orientation in the membrane region than in the non-membrane region.

c. *The spherical shell*: Working as in the previous case, the crack extension angle and the critical value of the applied pressure for crack propagation are determined for the hemispherical heads of the shell. When the crack exists in the membrane region relation (8.21) yields:

$$\sigma_\xi = \sigma_\theta = \sigma_1 = \sigma_2 = \frac{pR}{2h}. \tag{8.31}$$

This relation indicates that a material element in the membrane region is in a state of hydrostatic tension. Thus, the crack propagates in a self-similar manner ($\theta_0 = 0$) and the value of $\bar{S}_{cr}$ determined from relation (8.27) is equal to 6.40 (last columns of Tables 8.3 and 8.4).

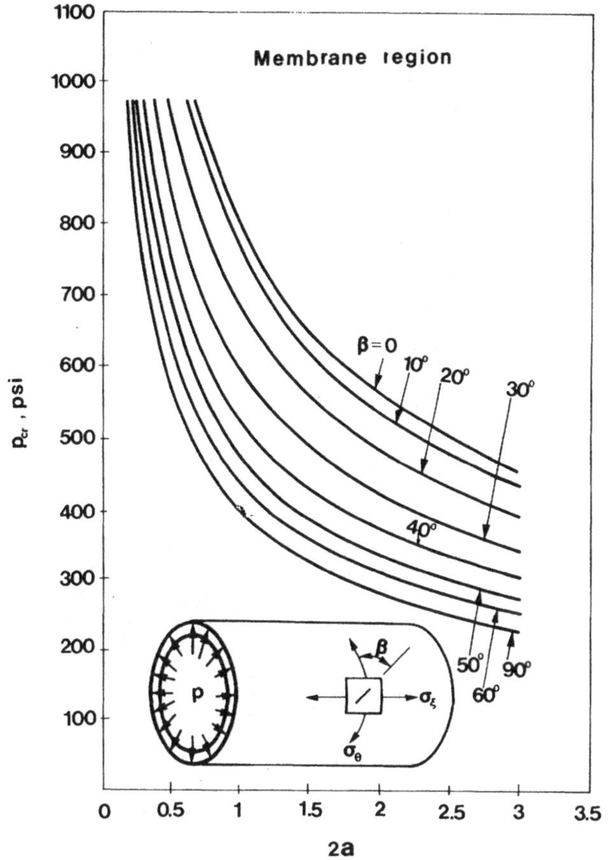

Figure 8.14. Variation of the critical pressure of a cylindrical shell versus the crack length for an element in the membrane region.

Table 8.3. Fracture angle versus crack angle and crack length for hemispherical shell ($z/h = 0.5$; $\nu = 0.3$)

$2a/h$ \ β	0°	10°	20°	30°	40°	50°	60°	70°	80°	90°
	Fracture angle, $-\theta_0$									
1	0	−1.02	−1.92	−2.61	−3.00	−3.04	−2.71	−2.03	−1.09	0
2	0	−0.47	−0.90	−1.22	−1.40	−1.41	−1.25	−0.94	−0.50	0
3	0	−0.16	−0.31	−0.42	−0.48	−0.48	−0.43	−0.32	−0.17	0
4	0	0.04	0.07	0.10	0.12	0.12	0.10	0.08	0.04	0
5	0	0.18	0.34	0.46	0.52	0.53	0.47	0.35	0.19	0
6	0	0.28	0.53	0.72	0.83	0.84	0.74	0.55	0.30	0
7	0	0.36	0.68	0.92	1.06	1.07	0.95	0.71	0.38	0
8	0	0.43	0.81	1.10	1.26	1.27	1.12	0.84	0.45	0
9	0	0.48	0.91	1.23	1.40	1.42	1.25	0.94	0.50	0
10	0	0.52	0.99	1.34	1.53	1.54	1.37	1.02	0.55	0
–	0	0	0	0	0	0	0	0	0	0

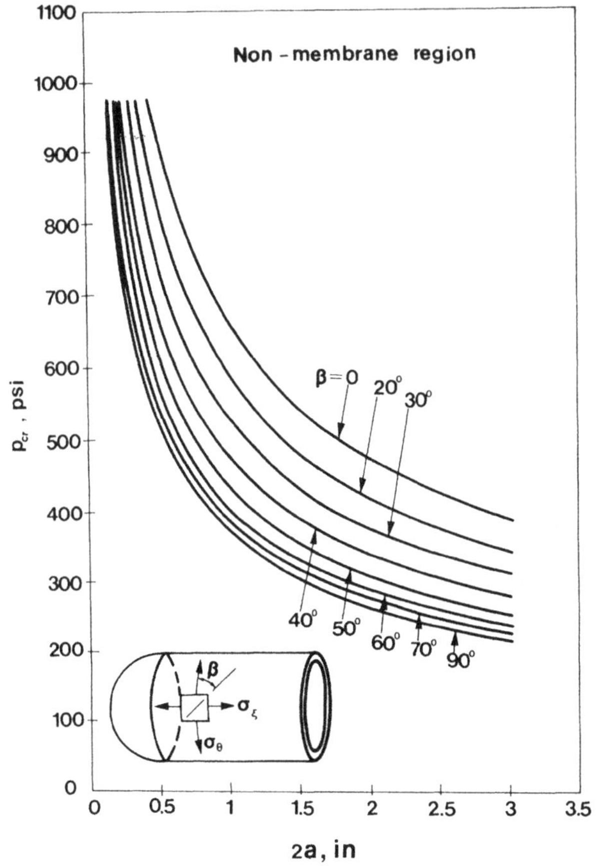

Figure 8.15. Variation of the critical pressure of a cylindrical shell versus the crack length for an element in the non-membrane region.

Table 8.4. Values of $\bar{S}_{cr}$ for hemispherical shell ($z/h = 0.5$; $\nu = 0.3$)

$2a/h$ \ β	0°	10°	20°	30°	40°	50°	60°	70°	80°	90°
1	8.28	8.24	8.15	8.00	7.83	7.64	7.46	7.32	7.23	7.19
2	8.04	8.02	7.93	7.81	7.66	7.50	7.35	7.23	7.16	7.13
3	7.94	7.92	7.84	7.72	7.59	7.44	7.30	7.19	7.12	7.10
4	7.87	7.84	7.77	7.67	7.54	7.40	7.27	7.17	7.10	7.08
5	7.82	7.80	7.73	7.63	7.50	7.37	7.25	7.15	7.09	7.07
6	7.79	7.77	7.71	7.61	7.49	7.36	7.24	7.14	7.08	7.06
7	7.77	7.74	7.68	7.58	7.47	7.34	7.23	7.13	7.07	7.05
8	7.73	7.71	7.65	7.56	7.44	7.32	7.21	7.12	7.06	7.04
9	7.71	7.69	7.63	7.54	7.43	7.31	7.20	7.11	7.05	7.03
10	7.69	7.67	7.61	7.52	7.42	7.30	7.19	7.10	7.05	7.03
–	6.40	6.40	6.40	6.40	6.40	6.40	6.40	6.40	6.40	6.40

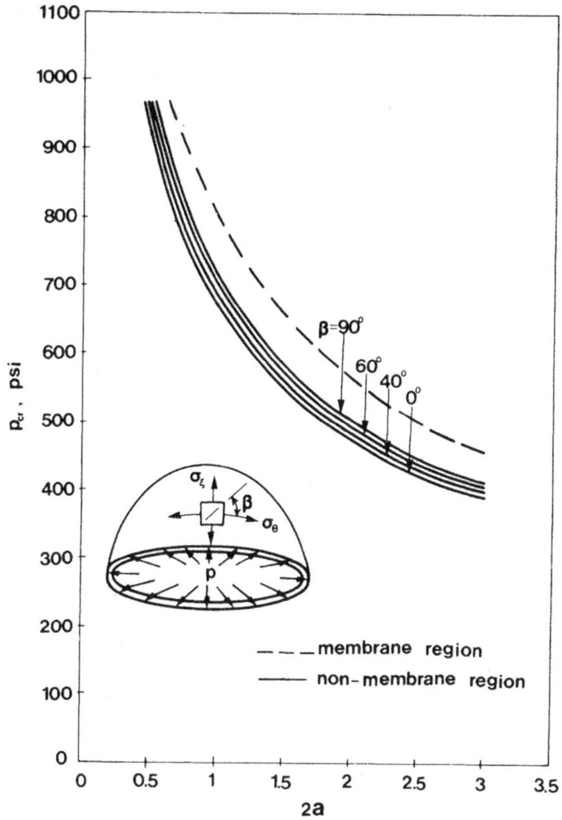

Figure 8.16. Variation of the critical pressure of the hemispherical heads of the shell of Figure 8.12 versus the crack length for an element in the membrane (dotted curve) or the non-membrane (solid curves) region.

When a crack of length $2a$ exists in the non-membrane region the values of $-\theta_0$ and $\bar{S}_{cr}$ are determined for a material element at a distance $\xi = 4.19$ in from the shell juncture (corresponding to an angle ϕ equal to 4° made by the radius aiming to the point and the equator). These values are shown in Tables 8.3 and 8.4 as a function of $2a/h$ and β.

Working as in the previous case of the cylindrical shell the values of the critical pressure p_{cr} for crack propagation are given by the following relations:

$$p_{cr} = \frac{1.4273}{\sqrt{a\bar{S}_{cr}}} \text{ ksi} \tag{8.32}$$

for the membrane region, and,

$$p_{cr} = \frac{1.3368}{\sqrt{a\bar{S}_{cr}}} \text{ ksi} \tag{8.33}$$

for the non-membrane region.

The variation of the quantity p_{cr} versus the crack length for the values $\beta = 0$, $40°$, $60°$ and $90°$ of the crack inclination angle is shown in Figure 8.16. The values of p_{cr} for the non-membrane region are presented by dotted lines. It is observed that p_{cr} for the non-membrane region is independent of crack inclination due to the hydrostatic state of stress developed in the shells. Figure 8.16 establishes the tendency of decreasing critical stresses with increasing crack lengths seen also in all cases studied previously. However, unlike results obtained for the cylindrical shell, p_{cr} increases with the crack inclination angle in the non-membrane region.

For more information on the results of this section the interested reader is referred to reference [15].

References

[1] Hartranft, R.J. and Sih, G.C., The use of eigenfunction expansions in the general solution of three-dimensional crack problems, *Journal of Mathematics and Mechanics*, 19, pp. 123–138 (1969).

[2] Williams, M.L., The bending stress distribution at the base of a stationary crack, *Journal of Applied Mechanics*, 28, pp. 78–82 (1961).

[3] Sih, G.C. and Rice, J.R., The bending of plates of dissimilar materials with cracks, *Journal of Applied Mechanics*, 31, pp. 477–482 (1964).

[4] Sih, G.C., Flexural problems of cracks in mixed media, *Proceedings of the First International Conference on Fracture*, Sendai, Japan, 1, pp. 391–409 (1965).

[5] Hartranft, R.J. and Sih, G.C., Effect of plate thickness on the bending stress distribution around through cracks, *Journal of Mathematics and Physics*, 47, pp. 276–281 (1968).

[6] Sih, G.C., A review of the three-dimensional stress problem for a cracked plate, *International Journal of Fracture Mechanics*, 7, pp. 39–61 (1971).

[7] Reissner, E., The effect of transverse shear deformation on the bending of elastic plates, *Journal of Applied Mechanics*, 67, pp. A69–77 (1945).

[8] Reissner, E., On bending of elastic plates, *Quarterly of Applied Mathematics*, 5, pp. 55–68 (1947).

[9] Knowles, J.K. and Wang, N.M., On the bending of an elastic plate containing a crack, *Journal of Mathematics and Physics*, 39, pp. 223–236 (1960).

[10] Hartranft, R.J. and Sih, G.C., An approximate three-dimensional theory of plates with application to crack problems, *International Journal of Engineering Science*, 8, pp. 711–729 (1970).

[11] Wang, N.M., Twisting of an elastic plate containing a crack, *International Journal of Fracture Mechanics*, 6, pp. 367–378 (1970).

[12] Sih, G.C., Strain energy density theory applied to plate bending problems, *Mechanics of Fracture 3*, edited by G.C. Sih, Noordhoff International Publishing, Leyden, pp. XVII–XLVIII (1977).

[13] Timoshenko, A. and Woinowsky-Kreiger, S., *Theory of Plates and Shells*, McGraw-Hill Book Company, Inc., New York, 1959.

[14] Au, N.N., Stresses in the shells under internal pressures, *Aerospace Corporation Report No. TDR-269(4304)-5*, January 10 (1964).

[15] Au, N.N. and Lin, S.R., Mixed mode fractures in space launch pressure vessels, *Proceedings of the U.S.–Greece conference on Mixed Mode Crack Propagation*, August 18–22, 1980, Athens, Greece, Editors G.C. Sih and P.S. Theocaris, Sijthoff and Noordhoff, pp. 55–76 (1981).

9

Three-dimensional crack problems

9.1. Introduction

The present chapter is devoted to the investigation of the growth characteristics of arbitrary cracks embedded in three-dimensional bodies. The general equations of the strain energy density criterion developed in chapter one are used to determine the critical loads for crack extension, as well as the new shape of the crack after propagation. The determination of the stress field along the crack front in three-dimensional crack problems is more complicated than in two-dimensional problems. A thorough study of three-dimensional crack problems was provided by Sih and his coworkers [1–6] who expressed the local stress field along the crack front in a form analogous to the two-dimensional case. Three stress intensity factors were used, all of which are independent of the local coordinates, depending only on the crack geometry, the form of loading and the location of the point along the crack border. This result is fundamental, in analyzing the fracture behavior of cracks, and provides uniform expressions for the local stresses under various geometrical and loading conditions where only the values of stress intensity factors differ. Kassir and Sih [7] analyzed three-dimensional crack problems under a variety of loads. They presented the stress field around embedded and external elliptical or circular cracks in a uniform form and gave the values of stress intensity factors under various loading forms. The variety of problems considered covers most of the corresponding cases of two-dimensional problems as presented by Sih [8], including single or interacting circular and elliptical cracks under normal, shear, torsional and thermal loads; semi-infinite plane cracks under a variety of loads; cracks along the interface of dissimilar materials; cracks in anisotropic and nonhomogeneous materials; and penny-shaped cracks under dynamic loads.

Application of the strain energy density criterion to all the above three-dimensional crack problems provides the critical loads and new shapes of cracks after propagation in a manner analogous to the first seven chapters of the book. However, it is not the intent of this book to provide the brittle fracture characteristics of the vast variety of fracture mechanics problems, but rather to give an idea of the usefulness of the strain energy density criterion for the solution of such problems. With this purpose in mind the present chapter will present only the case of an elliptical crack embedded in a solid loaded by a uniform inclined stress with respect to the crack plane. This case is analogous to the two-dimensional problem of an inclined crack under uniform stress studied in Chapter two (page 11). The critical stress for crack extension and the new shape of the crack after propagation are a function of the crack shape, the orientation of the applied stress and the Poisson's ratio of the material. Thus, solution of this fundamental problem is representative of the application of the strain energy density criterion to three-dimensional crack problems. Solutions to other cases can be found in the same manner as in the analogous cases of two-dimensional problems studied in the first seven chapters of the book.

9.2. The elliptical crack

a. *The local stress distribution along the crack front*: Let P_0 be a fixed point of the ellipse defining the crack front of an elliptical crack, and $\mathbf{n}_0$, $\mathbf{b}_0$ and $\mathbf{t}_0$ be the normal, binormal and tangent (unit) vectors at P_0 (Figure 9.1). Consider an arbitrary point P in the vicinity of P_0 and let r_0, θ_0, and ϕ_0 be the spherical coordinates of P. For this case and with respect to the system $P_0 n_0 b_0 t_0$ the stress components σ_n^0, σ_b^0, σ_t^0, τ_{nb}^0, τ_{nt}^0 and τ_{bt}^0 in a material element at point P are given by the following relations [6]:

$$\sigma_n^0 = \frac{k_1^0}{\sqrt{2r_0\cos\phi_0}}\cos\frac{\theta_0}{2}\left[1-\sin\frac{\theta_0}{2}\sin\frac{3\theta_0}{2}\right]-\frac{k_2^0}{\sqrt{2r_0\cos\phi_0}}\times$$
$$\times\sin\frac{\theta_0}{2}\left[2+\cos\frac{\theta_0}{2}\cos\frac{3\theta_0}{2}\right] \tag{9.1a}$$

$$\sigma_b^0 = \frac{k_1^0}{\sqrt{2r_0\cos\phi_0}}\cos\frac{\theta_0}{2}\left[1+\sin\frac{\theta_0}{2}\sin\frac{3\theta_0}{2}\right]+\frac{k_2^0}{\sqrt{2r_0\cos\phi_0}}\sin\frac{\theta_0}{2}\cos\frac{\theta_0}{2}\cos\frac{3\theta_0}{2} \tag{9.1b}$$

$$\sigma_t^0 = 2\nu\left[\frac{k_1^0}{\sqrt{2r_0\cos\phi_0}}\cos\frac{\theta_0}{2}-\frac{k_2^0}{\sqrt{2r_0\cos\phi_0}}\sin\frac{\theta_0}{2}\right] \tag{9.1c}$$

$$\tau_{nb}^0 = \frac{k_1^0}{\sqrt{2r_0\cos\phi_0}}\sin\frac{\theta_0}{2}\cos\frac{\theta_0}{2}\cos\frac{3\theta_0}{2}+\frac{k_2^0}{\sqrt{2r_0\cos\phi_0}}\cos\frac{\theta_0}{2}\left[1-\sin\frac{\theta_0}{2}\sin\frac{3\theta_0}{2}\right] \tag{9.1d}$$

$$\tau_{nt}^0 = -\frac{k_3^0}{\sqrt{2r_0\cos\phi_0}}\sin\frac{\theta_0}{2} \tag{9.1e}$$

$$\tau_{bt}^0 = \frac{k_3^0}{\sqrt{2r_0\cos\phi_0}}\cos\frac{\theta_0}{2}. \tag{9.1f}$$

where k_1^0, k_2^0 and k_3^0 are the stress intensity factors at point P_0 and ν is the Poisson's ratio of the material. The stress intensity factors $k_j^0 (j = 1, 2, 3)$ are defined such that they are independent of the coordinate variables (r_0, θ_0, ϕ_0) and depend only on the position of the point P_0 on the crack front, the nature of loading and the crack geometry.

Kassir and Sih [1] provided the stress field in the vicinity of a point along the periphery of an elliptical crack in a different, more complicated form with respect to another system of spherical coordinates and according to a different definition of the angles θ_0, ϕ_0.

Note in relations (9.1) that the $1/r^{1/2}$ singularity of stresses in the vicinity of the crack front usual in crack problems is still preserved. We also see that the stresses in an element in the normal $(\mathbf{n}_0\mathbf{b}_0)$ plane located at a distance $r = r_0\cos\phi_0$ from P_0 and at an

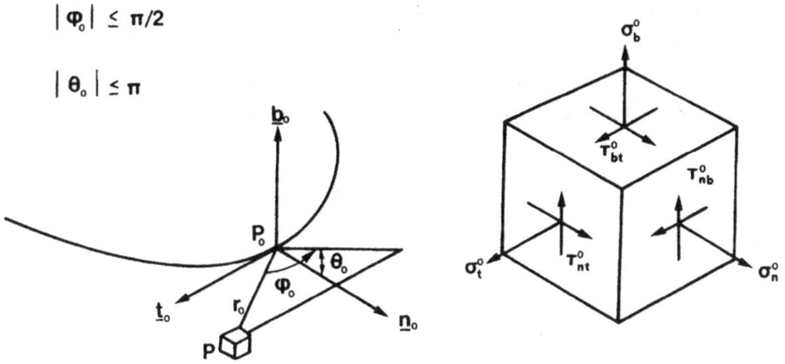

Figure 9.1. Crack front stresses in an arbitrary position.

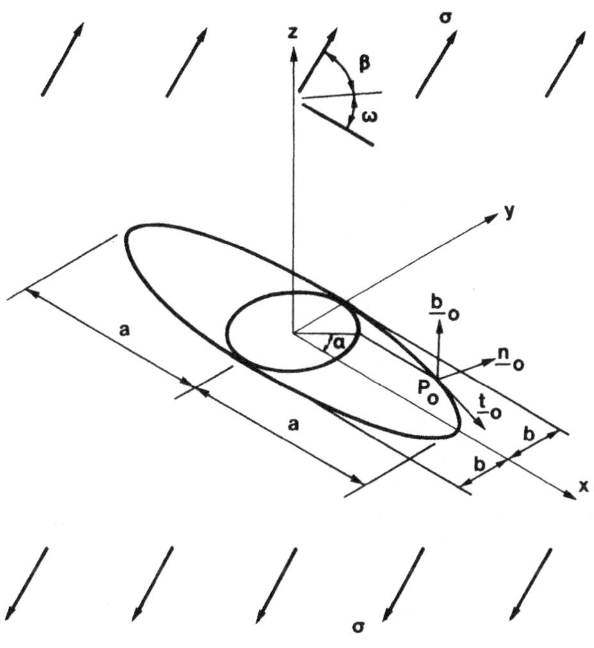

Figure 9.2. Elliptical crack at an angle to a field of uniform tension in an infinite material.

angle θ_0 from $\mathbf{n}_0$ are identical with those obtained in the two-dimensional problem (formulae (9.1) with $r = r_0 \cos \phi_0$ are identical with formulae (2.1)).

The stress field given by (9.1) is strictly valid for the elliptical crack. However, although not proven, it seems quite likely that similar results would be found for arbitrary cracks. The crack front may be a general curve in space described implicitly by its curvature and torsion and it will be assumed that the osculating plane of the crack front curve coincides with the tangent to the free surfaces of the crack at each point.

Let us now consider an infinite body subjected to a uniform uniaxial stress at infinity (Figure 9.2). An elliptical crack with semi-axes a and b which is embedded in this body lies in the xy-plane. The stress σ is applied at an angle β with reference to the crack plane, and at an angle ω with the xz-plane as shown in Figure 9.2. For this case the stress intensity factors k_1^0, k_2^0, and k_3^0 are given by [6]:

$$k_1^0 = \frac{\sigma\sqrt{(a_e)}}{E(k)} \sin^2\beta \tag{9.2a}$$

$$k_2^0 = \sigma\sqrt{(a_e)}k'k^2\frac{q}{a_e}\sin\beta\cos\beta\left[\frac{\sin\alpha\sin\omega}{(k^2+\nu k'^2)E(k)-\nu k'^2K(k)}+\right.$$
$$\left.+\frac{k'\cos\alpha\cos\omega}{(k^2-\nu)E(k)+\nu k'^2K(k)}\right] \tag{9.2b}$$

$$k_3^0 = (1-\nu)\sigma\sqrt{(a_e)}k'k^2\frac{a}{a_e}\sin\beta\cos\beta\left[-\frac{k'\cos\alpha\sin\omega}{(k^2+\nu k'^2)E(k)-\nu k'^2K(k)}+\right.$$
$$\left.+\frac{\sin\alpha\cos\omega}{(k^2-\nu)E(k)+\nu k'^2K(k)}\right] \tag{9.2c}$$

where $a_e = b\sqrt{(1-k^2\cos^2\alpha)}$

$$k' = \frac{b}{a}k = \sqrt{\left(1-\frac{b^2}{a^2}\right)}$$

and $K(k)$ and $E(k)$ are complete elliptic integrals of the first and second kinds, respectively.

It is observed that k_1^0, k_2^0 and k_3^0 depend on the angles of loading (β and ω), the crack diameter ratio (b/a), the position on the crack front (α), and the quantity $\sigma\sqrt{b}$.

b. *The strain energy density criterion:* As was previously pointed out, the local stress field along the crack border given by equation (9.1) is identical to the corresponding field for the two-dimensional problem if the material element is assumed to be on the plane normal to the crack front ($r = r_0\cos\phi$). Therefore, the expression for the strain energy density $\mathrm{d}W/\mathrm{d}V$ referred to each point along the crack periphery will be given by the same expression as in the two-dimensional case. Thus, we find from equations (2.2) to (2.4) for plane strain conditions that:

$$\frac{\mathrm{d}W}{\mathrm{d}V} = \frac{S}{r_0\cos\phi_0} + O(1) \tag{9.3}$$

where

$$S = a_{11}(k_1^0)^2 + 2a_{12}k_1^0k_2^0 + a_{22}(k_2^0)^2 + a_{33}(k_3^0)^2. \tag{9.4}$$

The coefficients a_{ij} ($i, j = 1, 2, 3$) are exactly the same as in equation (2.3) and are given by:

$$16\mu a_{11} = (3-4\nu-\cos\theta_0)(1+\cos\theta_0) \tag{9.5a}$$

$$16\mu a_{12} = 2\sin\theta_0(\cos\theta_0 - 1 + 2\nu) \tag{9.5b}$$

$$16\mu a_{22} = 4(1-\nu)(1-\cos\theta_0) + (3\cos\theta_0 - 1)(1+\cos\theta_0) \tag{9.5c}$$

$$16\mu a_{33} = 4. \tag{9.5d}$$

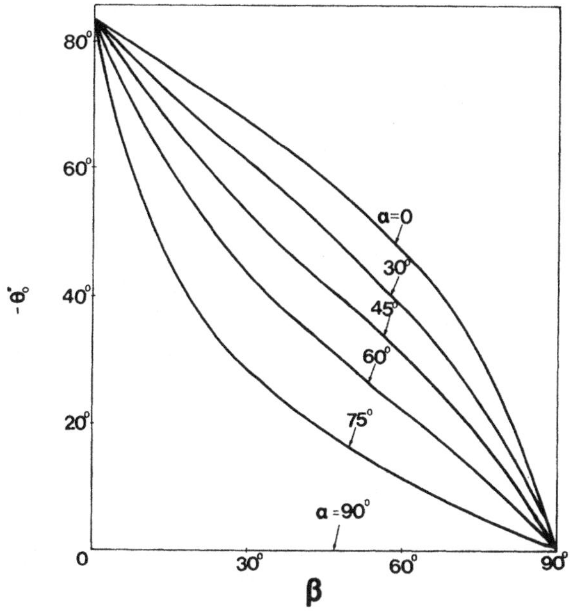

Figure 9.3. Variation of the fracture angle $-\theta_0^*$ versus the load inclination angle β for various positions (α) along the crack front. The applied stress is tensile and $b/a = 0.5$, $\nu = 0.33$ and $\omega = 0$.

Equations (9.3) to (9.5) give the strain energy density $\mathrm{d}W/\mathrm{d}V$ at each point along the crack front. According to the first hypothesis of the strain energy density criterion (Section 1.3, page 6) crack propagation from each point of the crack front is in the direction of the point with the minimum value of the strain energy density factor S' ($S' = (\mathrm{d}W/\mathrm{d}V)r_0 = S/\cos\phi_0$) as compared with other points of a spherically shaped core of radius r_0. Thus a line can be drawn from each point of the crack to the point of the sphere with the minimum value of S'. The aggregate of these lines defines the new crack surface.

The conditions of determining the minimum values of the strain energy density factor are expressed mathematically from relations (1.10) and (1.11). However, it is readily apparent from (9.3) that the minimum value of $\mathrm{d}W/\mathrm{d}V$ for the case under consideration occurs when

$$\phi_0^* = 0 \tag{9.6}$$

that is, the crack always propagates from each point of its front in the normal plane. The angle θ_0^* defining the position of the line in the normal plane that gives the new crack surface after propagation can be determined from the conditions (equations (1.10) and (1.11))

$$\frac{\partial S}{\partial \theta_0} = 0 \qquad \frac{\partial^2 S}{\partial \theta_0^2} > 0. \tag{9.7}$$

Having determined the angles $\phi_0^* = 0$ and θ_0^* which give the direction of crack propagation from each point of the crack front, the corresponding minimum values

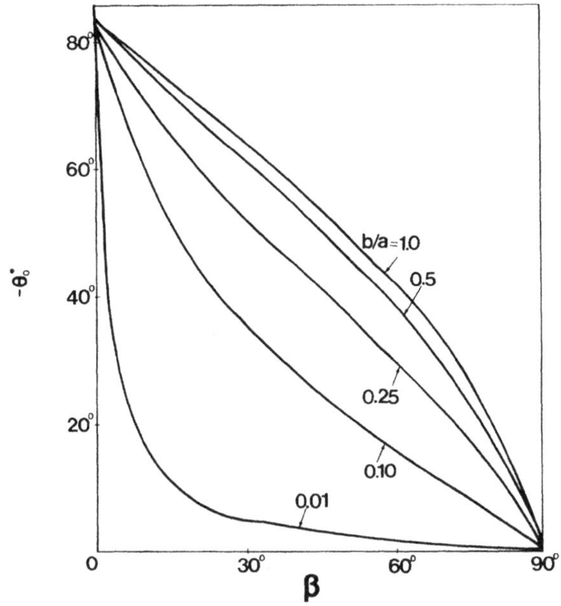

Figure 9.4. Variation of the fracture angle $-\theta_0^*$ versus the load inclination angle β for various values of the diameter ratio b/a of the crack. Curves of figure correspond to the position $\alpha = 30°$ along the crack front and $\nu = 0.33$ and $\omega = 0$.

$S(\phi_0^*, \theta_0^*)$ of S can be obtained. According to the second hypothesis of the strain energy density criterion (Section 1.3, page 6), crack propagation initiates when $S_{min} = S(\phi_0^*, \theta_0^*)$ reaches the maximum critical value S_{cr} tolerable by the material.

Thus, the crack starts to propagate from the point along its border with the maximum value of S_{min}. Crack propagation is a discontinuous rather than a continuous process. In determining the new crack surface immediately after propagation, the third hypothesis of the strain energy density criterion states that S_{min}/r_0 should remain constant along the crack front. This condition determines a different value of r_0 for each point of the crack front such that the ratio S_{min}/r_0 is kept constant. Thus, the entire fracture surface can be predicted.

c. *Fracture angles:* The values of the strain energy density factor S can be obtained as a function of the spherical angles ϕ and θ, the loading angles β and ω, the crack diameter ratio b/a, the position along the crack border α and the quantity $\sigma\sqrt{b}$ by substituting the values of the stress intensity factors k_1^0, k_2^0 and k_3^0 from equations (9.2) into equations (9.4) and (9.5). As was previously found, the minimum values of S correspond to $\phi_0^* = 0$, and the values of the angle θ_0^* are determined by relations (9.7). Negative angles θ_0 correspond to tensile loads, and positive angles θ_0 to compressive loads. Thus, the angle θ_0^* is determined to be a function of the quantities β, ω, b/a and α.

Figure 9.3 presents the variation of the fracture angle $-\theta_0^*$ versus the angle β for various positions along the crack front when the applied stress is tensile. The load is applied on the xz-plane ($\omega = 0$), $b/a = 0.5$ and $\nu = 0.33$. We see that when the applied

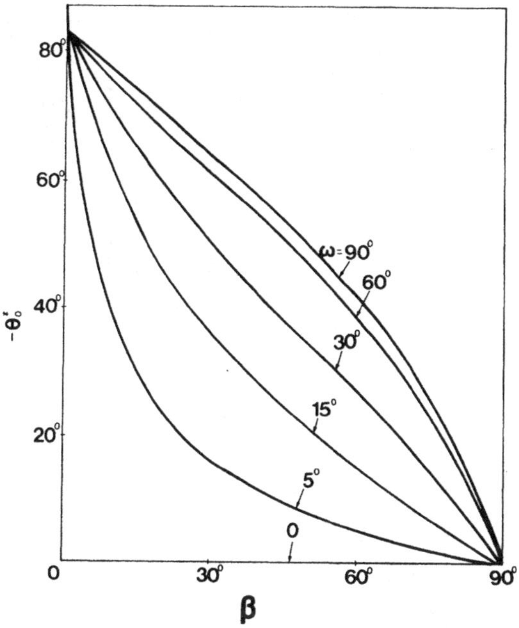

Figure 9.5. Variation of the fracture angle $-\theta_0^*$ versus the load inclination angle β for various values of the angle ω defining the direction of load with respect to the xz plane. Curves of figure correspond to the end of the minor axis of the ellipse ($\alpha = 90°$) and $b/a = 0.1$ and $\nu = 0.33$.

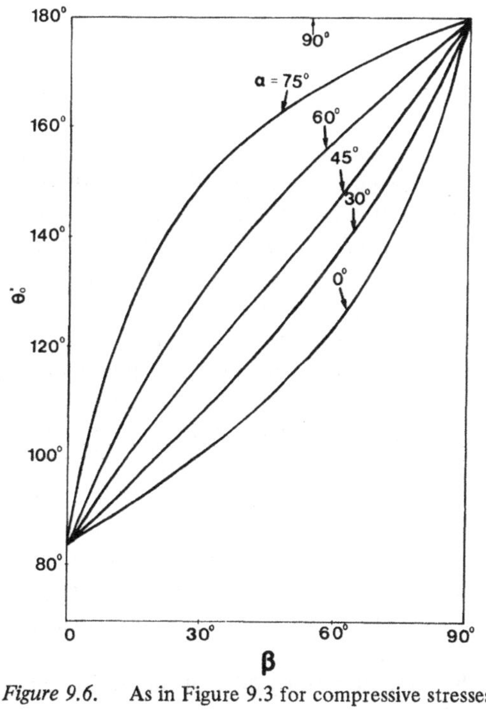

Figure 9.6. As in Figure 9.3 for compressive stresses.

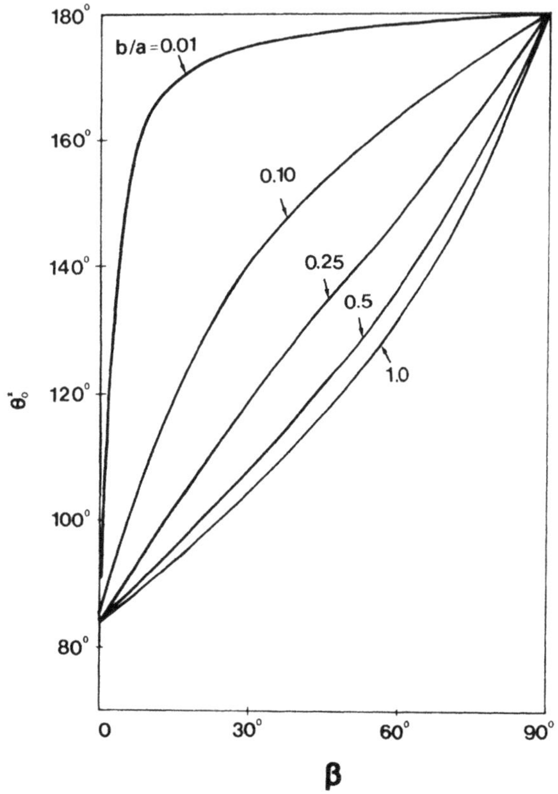

Figure 9.7. As in Figure 9.4 for compressive stresses.

stress is perpendicular to the crack plane ($\beta = 90°$), the crack grows from all points along its front in its own plane ($\theta_0^* = 0$). Furthermore, the ends of the minor axis ($\alpha = 90°$) grow in the plane of the crack. The tendency of the fracture angles $-\theta_0^*$ to decrease as the point on the crack border moves from the major to the minor axis of the crack is also apparent. Figure 9.4 establishes the influence of the shape of the ellipse (defined by the ratio of its diameters b/a) on the crack extension angle at the location $\alpha = 30°$. As in the previous figure, ω and ν take the values 0 and 0.33 respectively. As expected the fracture angles for longer ellipses, which approach two-dimensional cracks, tend to zero. Finally, Figure 9.5 gives fracture angles $-\theta_0^*$ at the end of the minor axis ($\alpha = 90°$) for several values of the angle ω in between the direction of loading and the plane xz, for $b/a = 0.1$ and $\nu = 0.33$.

In Figures 9.3 to 9.5 the applied stress is tensile, resulting in negative fracture angles, $-\theta_0^*$. Figures 9.6, 9.7 and 9.8 present analogous results as Figures 9.3, 9.4 and 9.5 for compressive stresses. We see that the fracture angle θ_0^* in Figures 9.6, 9.7 and 9.8 is positive. From Figures 9.6, 9.7 and 9.8 observations regarding the dependence of the fracture angle θ_0^* on the parameters β, ω, b/a, α and ν analogous to those of Figures 9.3, 9.4 and 9.5 can be made.

d. *Critical loads*: The minimum values S_{min} of the strain energy density factor S are

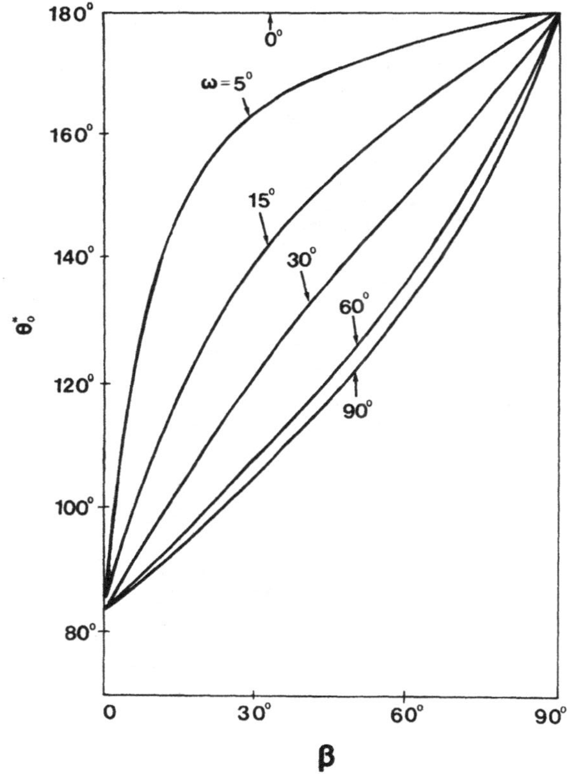

Figure 9.8. As in Figure 9.5 for compressive stresses.

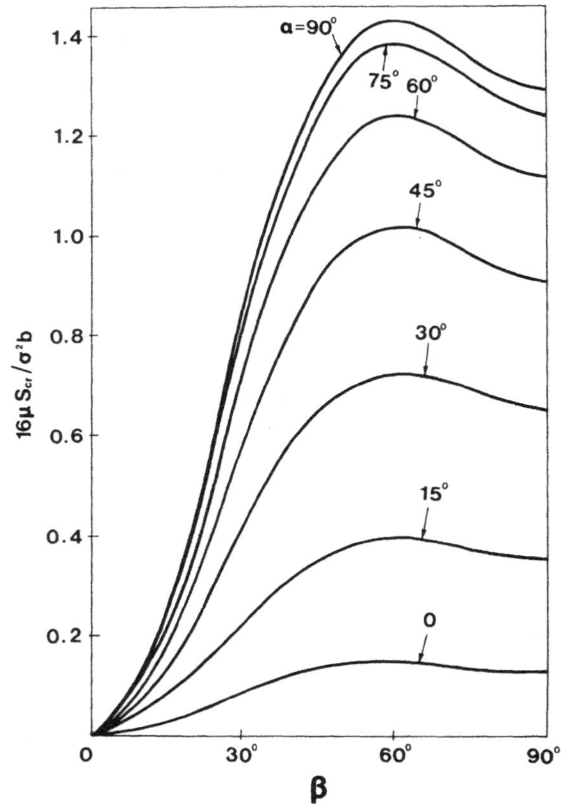

Figure 9.9. Variation of the non-dimensional quantity $16\mu S_{\mathrm{cr}}/\sigma^2 b$ versus β for various locations (α) along the crack front. $b/a = 0.1$, $\nu = 0.33$, $\omega = 0$.

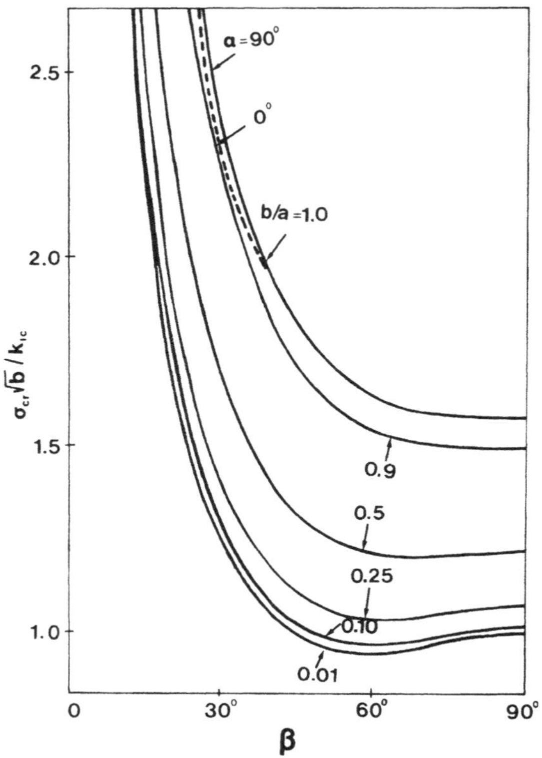

Figure 9.10. Variation of the normalized critical fracture stress $\sigma_{cr}\sqrt{b}/k_{1c}$ versus the load inclination angle β for various values of the diameter ratio b/a of the crack. Critical loads occur at $\alpha = 90°$ except for dashed curve where they occur at $\alpha = 0$, $\nu = 0.33$ and $\omega = 0$.

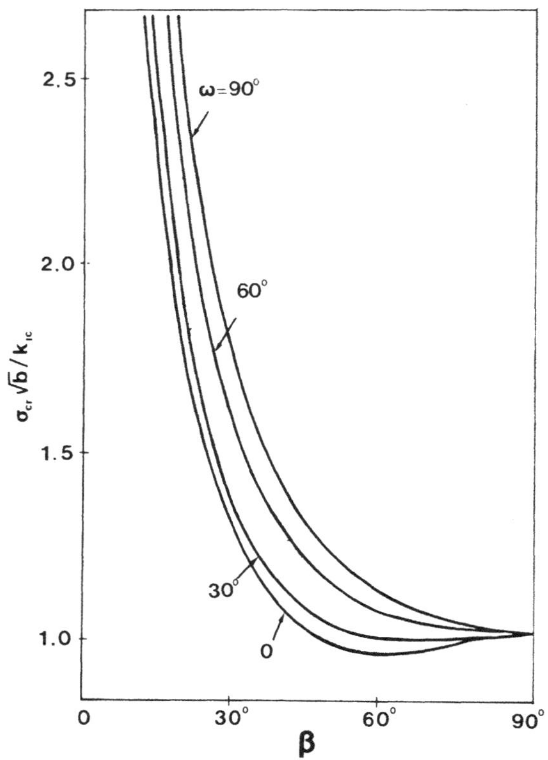

Figure 9.11. Variation of the normalized critical fracture stress $\sigma_{cr}\sqrt{b}/k_{1c}$ versus the load inclination angle β for various values of the angle ω that the plane of the stress forms with the xz-plane. Critical loads occur at $\alpha = 90°$, $b/a = 0.1$ and $\omega = 0.33$.

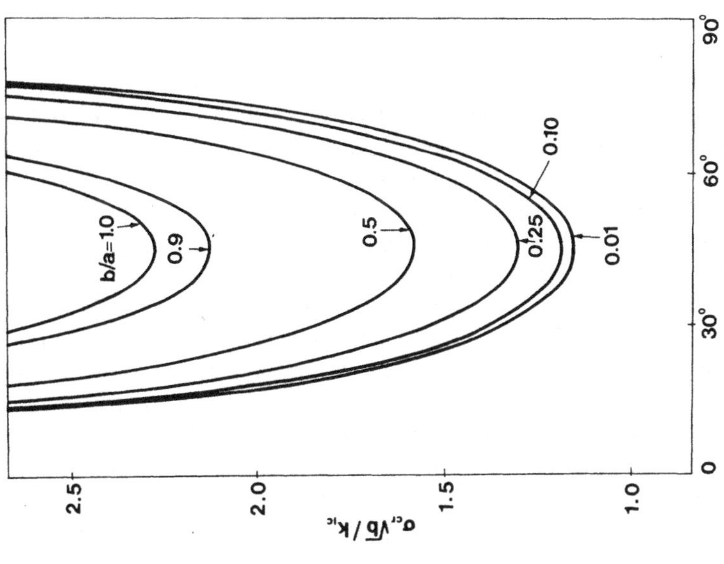

Figure 9.12. As in Figure 9.10 for compressive stresses.

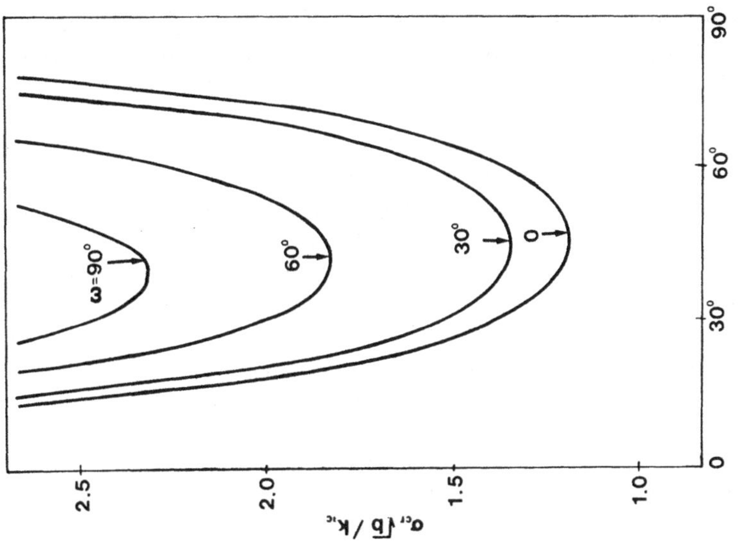

Figure 9.13. As in Figure 9.11 for compressive stresses.

obtained by introducing the values of the fracture angles $-\theta_0^*$ and $\phi_0 = 0$ into equations (9.4) and (9.5). Figure 9.9 presents the variation of the non-dimensional quantity $16\mu S_{cr}/\sigma^2 b$ versus β for various positions along the crack border (defined by the angle α) and for $\omega = 0$, $b/a = 0.1$, and $\omega = 0.3$. The applied stress is tensile. It is observed that S_{min} takes its largest value at the end point of the minor axis of the ellipse ($\alpha = 90°$) and therefore crack extension initiates from that point.

The value of the critical applied stress σ_{cr} for crack extension can be determined by equating the values of S_{min} with the critical value of the strain energy density factor, S_{cr}, (which is a material constant). Figure 9.10 presents the variation of the non-dimensional quantity $\sigma_{cr}\sqrt{b}/k_{1c}$ versus the load inclination angle β for various values of the diameter ratio a/b of the ellipse under tensile applied loads. k_{1c} is the critical value of the strain energy density factor connected with S_{cr} through relation (2.16). The Poisson's ratio ν takes the value 0.33, and $\omega = 0$. As previously indicated, crack extension generally initiates from the end of the minor axis of the ellipse. However, this does not always hold true for nearly penny-shaped cracks ($a/b \simeq 1$). Thus, in Figure 9.10 two curves for the penny-shaped crack are shown, corresponding to the ends of the minor ($\alpha = 90°$, solid curve) and the major ($\alpha = 0$, dotted curve) axis of the ellipse. For $\beta \leqslant 60°$ the values σ_{cr} of the applied stress σ for crack extension from the end of the major axis are smaller than those required to initiate fracture from the end of the minor axis of the ellipse. Thus, for $\beta \leqslant 60°$ crack extension takes place from the end of the major axis, while for the remaining values of β, crack extension will initiate from the end of the minor axis. The trend of decreasing critical stresses as the ellipse becomes increasingly slender (decreasing the value of b/a) is illustrated in Figure 9.10. The effect of the load inclination angle ω on the fracture stress σ_{cr} is shown in Figure 9.11. It is observed that the stress σ_{cr} decreases as ω decreases, and takes its minimum value when the applied stress lies in the xz-plane.

For compressive applied stresses, the values of the critical stress for crack extension corresponding to the geometrical and loading conditions of Figures 9.10 and 9.11 are shown in Figures 9.12 and 9.13 respectively. The curves of Figures 9.12 and 9.13 are symmetric with respect to $\beta = 45°$, at which σ_{cr} becomes minimum and tend to infinity as β tends to 0 or 90°. Otherwise the same trends regarding the dependence of σ_{cr} on b/a and ω as in Figures 9.10 and 9.11 are observed in Figures 9.12 and 9.13.

e. *Post fracture crack shape:* As previously determined, crack extension always takes place from each point along its front on the normal plane, and the fracture angles θ_0^* were calculated from the minimum values of the strain energy density factor. Thus, a line drawn from each point on the crack front in the normal plane at an angle θ_0^* with respect to the crack plane indicates the directions in which the points of the crack front would move after propagation. The third hypothesis of the strain energy density criterion which states that S_{min}/r_0 remains constant along the crack front is used to determine the new fracture surface after crack propagation. Thus, the values of r_0 at all points of the crack periphery are calculated from the previously obtained minimum values S_{min} while specifying the length r_0 of the crack extension at a particular point of the crack. The resulting initial segment of the new crack surface around an elliptical crack of $b/a = 0.5$ and $\beta = 45°$ with the applied stress in the xz-plane ($\omega = 0$) is shown in Figure 9.14 when the stress is tensile. The corresponding shape for compressive stresses is shown in Figure 9.15. Unlike to the tension case, however, the

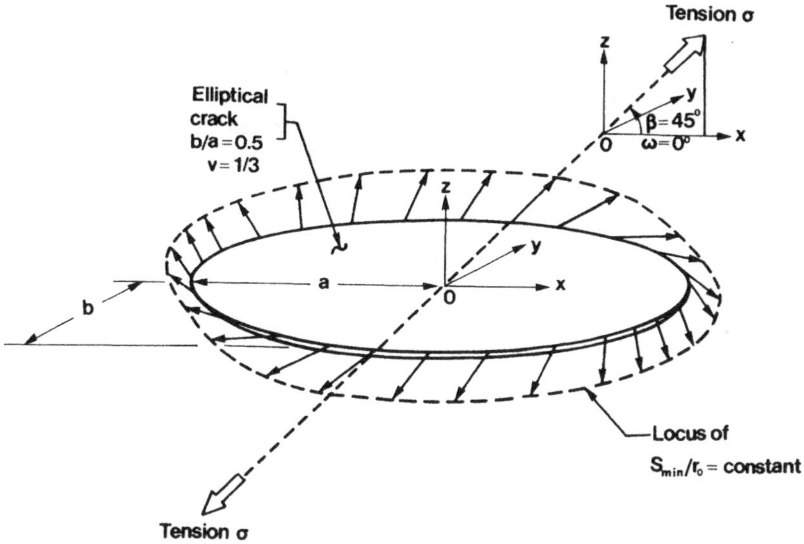

Figure 9.14. Initial fracture surface of an elliptical crack ($b/a = 0.5$) under tension, $\beta = 45°$ and $\omega = 0°$.

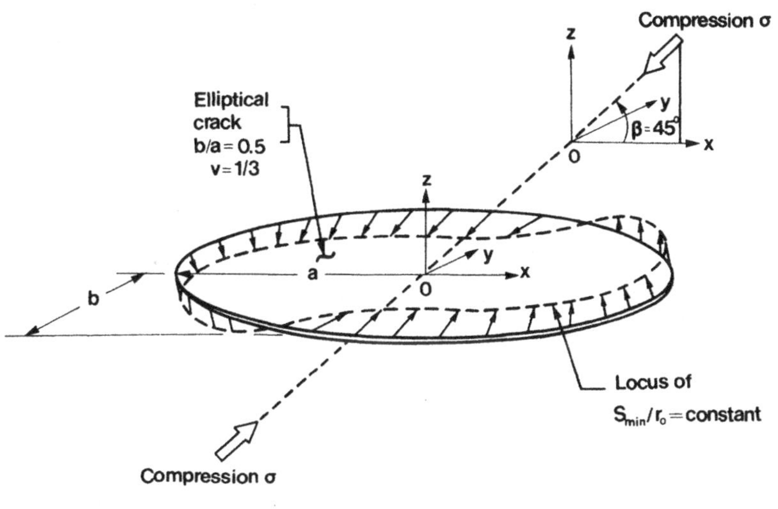

Figure 9.15. Initial fracture surface of an elliptical crack ($b/a = 0.5$) under compression, $\beta = 45°$ and $\omega = 0°$.

initial fracture surface under compressive stress is warped considerably and slants toward the direction of loading.

For more information on the results of this section the interested reader is referred to references [5, 6].

References

[1] Kassir, M.K. and Sih, G.C., Three-dimensional stress distribution around an elliptical crack under arbitrary loading, *Journal of Applied Mechanics*, 33, pp. 601–611 (1966).

[2] Hartranft, R.J. and Sih, G.C., The use of eigenfunction expansions in the general solution of three-dimensional crack problems, Journal of *Mathematics and Mechanics*, 19, pp. 123–138 (1969).

[3] Sih, G.C. and Cha, B.C.K., A fracture criterion for three-dimensional crack problems, *Engineering Fracture Mechanics*, 6, pp. 699–723 (1974).

[4] Hartranft, R.J. and Sih, G.C., Three-dimensional growth characteristics of a plane crack subjected to concentrated forces, *Journal of Applied Mechanics*, 47, pp. 808–809 (1974).

[5] Sih, G.C., A three-dimensional strain energy density factor theory of crack propagation, *Mechanics of Fracture 2*, edited by G.C. Sih, Noordhoff International Publishing, Leyden, pp. XV–LIII (1975).

[6] Hartranft, R.J. and Sih, G.C., Stress singularity for a crack with an arbitrarily curved front, *Engineering Fracture Mechanics*, 9, pp. 705–718 (1977).

[7] Kassir, M.K. and Sih, G.C., Three-dimensional crack problems, *Mechanics of Fracture 2*, edited by G.C. Sih, Noordhoff International Publishing, Leyden (1975).

[8] Sih, G.C., Handbook of Stress Intensity Factors, Lehigh University (1973).

Author index

Subject index

GPSR Compliance

The European Union's (EU) General Product Safety Regulation (GPSR) is a set of rules that requires consumer products to be safe and our obligations to ensure this.

If you have any concerns about our products, you can contact us on ProductSafety@springernature.com

In case Publisher is established outside the EU, the EU authorized representative is:

Springer Nature Customer Service Center GmbH
Europaplatz 3
69115 Heidelberg, Germany

Batch number: 10627713

Printed by Printforce, the Netherlands